FIGHTER DRAWDOWN DYNAMICS
Effects on Aircrew Inventories

William W. Taylor • James H. Bigelow • John A. Ausink

Prepared for the United States Air Force

Approved for public release; distribution unlimited

PROJECT AIR FORCE

The research described in this report was sponsored by the United States Air Force under Contract FA7014-06-C-0001. Further information may be obtained from the Strategic Planning Division, Directorate of Plans, Hq USAF.

Library of Congress Cataloging-in-Publication Data is available for this publication.

ISBN 978-0-8330-4695-6

Published 2009 by the RAND Corporation
1776 Main Street, P.O. Box 2138, Santa Monica, CA 90407-2138
1200 South Hayes Street, Arlington, VA 22202-5050
4570 Fifth Avenue, Suite 600, Pittsburgh, PA 15213-2665
RAND URL: http://www.rand.org/
To order RAND documents or to obtain additional information, contact
Distribution Services: Telephone: (310) 451-7002;
Fax: (310) 451-6915; Email: order@rand.org

Preface

This monograph documents recent (2005–2008) RAND Corporation research on combat air force aircrew management problems resulting from the competing goals of producing sufficient experienced combat pilots and operating within the constraints of force structure reductions.

The monograph summarizes the supply-and-demand problems the fighter force has faced over the past decade, summarizes the decisions made in an attempt to solve them, and describes the RAND dynamic simulation model used to help the Air Force better understand the consequences for fighter units of attempting to maintain high fighter-pilot production levels while the fighter infrastructure is declining. We use the model to show the potential unsatisfactory consequences of some aircrew management polices that were introduced from 2005 to 2008 and then to develop policy options that would enable the Air Force to maintain a healthy fighter pilot force and address the added complications of the rising demand for fighter pilots in various nonflying positions.

This research is part of a multiyear study, "Rated Force Management," sponsored by the Air Force Deputy Chief of Staff for Air, Space, and Information Operations, Plans and Requirements (AF/A3/5). The study was conducted within the Manpower, Personnel, and Training Program of RAND Project AIR FORCE. This monograph is designed to help aircrew managers and analysts support senior Air Force policymakers in developing policies that will maintain a healthy fighter pilot force.

iv Fighter Drawdown Dynamics: Effects on Aircrew Inventories

Readers may also be interested in the following related RAND documents:

- *The Air Force Pilot Shortage: A Crisis for Operational Units?* by William W. Taylor, S. Craig Moore, and C. Robert Roll, Jr., MR-1204-AF, 2000.
- *Absorbing Air Force Fighter Pilots: Parameters, Problems, and Policy Options,* by William W. Taylor, James H. Bigelow, S. Craig Moore, Leslie Wickman, Brent Thomas, and Richard Marken, MR-1550-AF, 2002.
- *Absorbing and Developing Qualified Fighter Pilots: The Role of the Advanced Simulator,* by Richard S. Marken, William W. Taylor, John A. Ausink, Lawrence M. Hanser, C. R. Anderegg, and Leslie Wickman, MG-597-AF, 2007.

RAND Project AIR FORCE

RAND Project AIR FORCE (PAF), a division of the RAND Corporation, is the U.S. Air Force's federally funded research and development center for studies and analyses. PAF provides the Air Force with independent analyses of policy alternatives affecting the development, employment, combat readiness, and support of current and future aerospace forces. Research is conducted in four programs: Force Modernization and Employment; Manpower, Personnel, and Training; Resource Management; and Strategy and Doctrine.

Additional information about PAF is available on our Web site: http://www.rand.org/paf

Contents

Figures

Tables

Summary

The Air Force faces an increasing demand for personnel with pilot skills, a demand driven by the expanding number and size of various staffs (such as those of air operations centers) and an increased demand for operators of unmanned aircraft systems (UASs), who, by Air Force policy, must be pilots.[1] At the same time, the Air Force faces a declining ability to produce pilots (particularly fighter pilots) because its aircraft inventory is decreasing: By 2016, there will be fewer than 1,000 fighter aircraft in the total Air Force inventory (active, Guard and Reserve)—only 32 percent of the number in 1989. With fewer aircraft, it is difficult for all pilots to fly enough to maintain their combat skills, and it is particularly difficult for new pilots to gain enough experience in their first flying tour to be prepared for follow-on nonflying and flying positions (pp. 54–60). This has made aircrew management—the process of maintaining an aircrew force of sufficient size and readiness to accomplish the Air Force's current and forecast mission—particularly difficult, and without changes to current management policies, the Air Force will be unable to fill important flying and staff positions.

Our primary conclusion is that, to maintain the health of fighter units, the number of new pilots entering them must be reduced, ultimately to below 200 per year by 2016. Overabsorption decreases the number of monthly sorties inexperienced pilots can fly, lowers the average experience level of fighter units, makes it difficult or impossible

---

[1] Because *UAS* is the term that refers to the equipment the pilots operate directly, it is the term commonly used among military pilots and in this monograph to refer to skills, equipment, and operations related to *unmanned aerial vehicles*.

for new pilots to become experienced in an initial three-year tour, and increases the amount of time a pilot must wait between a first flying tour and an opportunity for a second one, thus slowing the development of the background and experience elements needed to make the pilot useful in staff or supervisory positions. All these circumstances can lead to a general degradation of pilot skills and combat capability.

The current situation in fighter units is the result of several decisions made during the Cold War drawdown in Air Force structure (p. 7). The drawdown resulted in a surplus of pilots, and the Air Force responded to the surplus by decreasing pilot production in the early 1990s without encouraging the departure of pilots already in the inventory (pp. 7–8). The Air Force also closed down one of its pilot training bases. By 1996 the fighter force had declined enough to require increasing pilot production, but the earlier base closure made the desired increase (almost doubling fighter-pilot production) difficult. In addition, the low production of the early 1990s meant that a shortage of pilots in specific skill groups and grades had developed (p. 7). Partly to address the shortage, but also to meet increasing demands for pilots in staff positions, a series of four-star level summits from 1996 to 2003 established annual pilot-production goals that remained in force through 2007: 1,100 total pilots, of which 330 were to be fighter pilots (pp. 8–10).

In 2005, RAND used a dynamic mathematical model of fighter pilot absorption capacity (the process of turning a new pilot into an "experienced" one) to show that attempting to achieve the established fighter-pilot production goals with the projected aircraft inventory would severely impair the ability of fighter units to maintain combat capability and provide the training necessary to produce experienced pilots (pp. 42–45). Units would have more personnel than they were authorized for the number of aircraft they had; inexperienced pilots would not be able to fly the number of sorties they needed to each month; and new fighter pilots would complete their first three-year tours without receiving the experience they needed for follow-on flying assignments or staff tours (p. 45).

These results, along with the recognition that other policy changes would affect the health of fighter units, led to the establishment of the

Transformational Aircrew Management Initiatives for the 21st Century (TAMI 21) working group in October 2006. Its goal was to find solutions to various aircrew management problems and present them to senior leadership. The RAND dynamic model enabled rapid analysis of the consequences of policy solutions that the TAMI 21 group discussed. The group's primary conclusion was that the flow of new pilots into fighter units must be reduced to a level at which units could absorb them—about 200 new fighter pilots per year. Using RAND modeling, the group also showed, however, that the Air Force could still maintain an inventory of personnel with the desired pilot skills (and satisfy emerging UAS requirements) if it implemented several policy initiatives (pp. 52–67).

After a four-star level aircrew management conference in March 2007, the Air Force implemented a much-revised version of the TAMI 21 recommendations. The most important decision was an attempt to simultaneously meet increased demands for UAS pilots and decrease the overmanning of fighter units by transferring a limited number of inexperienced pilots from fighter units to UAS aircraft before they had completed their first three-year tour. Unfortunately, RAND modeling showed that, while this decision led to short-term improvements in F-16 and F-15C fighter units, it would still be necessary to make long-term reductions in the flow of new pilots into these units to keep them from becoming "broken" by 2013.[2] Using the dynamic model, RAND was able to define one option for pilot-production reductions that would allow the F-15C and F-16 units to remain healthy through 2016 (pp. 72–76).

As 2008 came to a close, the Air Force faced a large increase in the demand for UAS operators: An April 2008 document shows the demand increasing from 458 in FY 2008 to 1,060 by FY 2013 (p. 67). Emerging air operations center and numbered air force staff requirements could increase pilot requirements by another 1,000 (Carney, 2008)—and this while some major command staffs are already able to

---

[2] "Broken" is a technical term defined in Chapter Two.

fill fewer than 75 percent of their rated staff billets.[3] At the same time, the fighter aircraft inventory continues its decline, and the replacement of F-16s with Joint Strike Fighter aircraft may be delayed.[4]

Since RAND's modeling has shown that, under current conditions, the flow of pilots into active-duty fighter units must be reduced, the Air Force faces both a supply and a demand problem for people with fighter pilot skills. The demand problem can be addressed in at least two ways:

1. Reduce demand by closely examining emerging staff requirements and eliminating those that are not actually required.

2. For positions that have been validated and that do appear to require personnel with fighter pilot experience, explore the full potential of other available manning alternatives, such as appropriately developed personnel currently affiliated with the Guard and Reserve, career enlisted aviators, and civilians with prior military experience.

The supply problem can be addressed in at least two ways:

1. Increase the supply of fighter pilots by effectively using the total force aircraft inventory (active, Air National Guard, and Air Force Reserve force structure) to absorb and develop new active pilots.

2. Increase the supply of UAS pilots by establishing an independent, self-sustaining UAS career field. The current requirement that UAS operators who are fighter pilots must be able to return to a fighter unit is unmanageable—there are not enough fighter aircraft to allow it. Creating a UAS career field (and not treating it as an air liaison–forward air control–Air Education and Training Command [ALFA] tour) will decrease stress on fighter units and increase the supply of UAS operators. A short-term

---

[3]   Email communication from Air Force Personnel Center, Deputy Personnel Assignment Operation (AFPC/DPAO), February 5, 2008.

[4]   The Air Staff's Distribution Plan Version 4.1, May 2007 does not show anyone filling Joint Strike Fighter cockpits as late as FY 2016.

solution here might involve sending specialized undergraduate pilot training graduates to UAS tours. A longer-term solution might involve changing the requirements for UAS operators (requiring, for example, only the first half of specialized undergraduate pilot training to become one).

It is critical that the Air Force curtail the flow of new pilots into active fighter units to avoid exceeding the current absorption constraints of the training system.[5] While, in the short term, reducing this flow could lead to shortfalls for some Air Force needs, the supply and demand options listed above will still allow the Air Force to meet expanding needs in other areas. Failing to reduce the flow will, in the near future, damage the combat capability of fighter units.

---

[5] Increasing the aircraft inventory would, of course, solve the absorption problem, but this solution is highly unlikely. Allowing simulator hours to provide more credit toward experience requirements and RAP requirements has the potential to increase the absorption capacity of units, but this requires investments in the improvement of simulator infrastructure and capabilities.

Acknowledgments

This research owes its existence to the knowledge and cooperation of individuals from a number of Air Force agencies. We appreciate the ongoing support of our principal project sponsor, Lt Gen (now Gen) Carrol Chandler, who ensured that we remained engaged in the analytic process, despite the serious issues that our analyses generated for the Air Force leadership. We express our special thanks for the continuing counsel of long-term Air Force aircrew management experts, James "Robbie" Robinson of the Resource Requirements branch in AETC's directorate of Intelligence, Air, Space and Information Operations (A3R); Craig Vara of the Force Management branch in Air Mobility Command headquarters (AMC/A3TF); Ed "Buck" Tucker of the Flight Management Branch in Air Combat Command (ACC/A3TB), Lt Col Kent Barker of the Rated Management branch of the National Guard Bureau's Force Management Division (ANG/A1FF), and C. J. Ingram from the Aircrew Management branch under Operational Training at HQ Air Force (AF/A3O-AT), who provided continuity and historical perspective throughout the Transformational Aircrew Management Initiatives for the 21st Century (TAMI 21) effort.

Other TAMI 21 Task Force members who deserve special thanks for their contributions include Lt Col (now Col) Frank Van Horn, who, as the Air Force A3O-AT branch chief, exhibited remarkable leadership and patience throughout the TAMI 21 process; Lt Col (now Col) Michael Hornitschek, who as branch chief for Rated Force Policy (AF/A1PPR) in the Directorate for Manpower, Personnel and Services provided rated personnel policy perspective; and Col William "Woody"

Watkins, Director of Operational Assignments at the Air Force Personnel Center (AFPC), who contributed thoughtful commentary and very useful information throughout our discussions.

Several others have continued to contribute very useful information and data. They include Thomas (Tom) Winslow, John Wigle, Maj Mike Rider of AF/A3O-AT, and Maj Russel Garner of the Combat Air Force Assignments branch (AFPC/DPAOC) at AFPC. Other important Air Staff contributors were Col Chuck Armentrout, Chief of the Military Force Policy Division (AF/A1PP) in the Air Force's Directorate for Manpower, Personnel and Services, and Lt Col (now Col) Kip Turain, of the Rated Force Policy branch (AF/A1PPR). Additional important information, discussion, and thoughtful review were contributed by a number of ACC staff members, including Col Eric "PJ" Best, Chief of ACC's Flight Operations Division (ACC/A3T); Col John Hart, Reserve Advisor to the Commander of ACC (ACC/CR); Col Joe Speckhart, Reserve Advisor to ACC's Director of Plans and Programs (ACC/A5H); Maj Barley Baldwin and Maj (now Lt Col) Chris Davis, F-15C and F-16 functional area managers, respectively, in ACC's Operations and Training branch (ACC/A3TO) (of which Lt Col Davis is now chief); and Joe Shirey and Chuck Higgins of ACC/A3EZ.

We thank our RAND colleagues Ryan Henry, Harry Thie and Louis "Kip" Miller for reviewing the document and providing excellent suggestions for improving the presentation of the material.

Finally, we would like to express our appreciation for the continuing support, leadership, and encouragement that we received from our RAND colleague and former supervisor, Natalie Crawford.

Abbreviations

ACC	Air Combat Command
ACC/A3EZ	Air Combat Command, Air and Space Operations, Strategic Planning Group
ACC/A3T	Air Combat Command, Air and Space Operations, Flight Operations Division [formerly ACC/DOT]
ACC/A3TB	Air Combat Command, Air and Space Operations, Flight Management Branch
ACC/A3TO	Air Combat Command, Air and Space Operations, Fight Operations and Training Branch
ACC/CR	Air Combat Command, Reserve Command Advisor
AETC	Air Education and Training Command
AETC/A3R	the resource requirements branch in AETC's directorate of Intelligence, Air, Space and Information Operations
AF/A1	Air Force, Manpower, Personnel and Services [formerly AF/DP]
AF/A1PP	Air Force, Manpower, Personnel and Services, Force Management Policy, Military Force Policy Division

AF/A1PPR	Air Force, Manpower, Personnel and Services, Force Management Policy, Military Force Policy Division, Rated Force Policy Branch
AF/A3/5	Air Force, Operations, Plans and Requirements
AF/A3O-AT	Operational Training Division of Air Force, Operations
AF/A8P	Air Force, Deputy Chief of Staff for Strategic Plans and Programs, Directorate of Programs [formerly AF/XPP]
AF/A8PE	Air Force, Directorate of Programs, Program Integration Division
AFB	air force base
AFPC	Air Force Personnel Center
AFPC/DPAO	Air Force Personnel Center, Operational Assignments
AFRC	Air Force Reserve Command
AFSO-21	Air Force Smart Operations for the 21st Century
AFSOC	Air Force Special Operations Command
ALFA	ALO, FAC [forward air control], and AETC
ALO	air liaison officer
AMC	Air Mobility Command
AMC/A3TF	Force Management Branch in Air Mobility Command headquarters
ANG	Air National Guard
AOC	air operations center
API	aircrew position indicator
ASD	average sortie duration

B-course	basic course
BMC	basic mission capable
BRAC	Base Realignment and Closure
CAF	combat air forces
CEA	career enlisted aviator
CMR	combat mission ready
CSAF	Chief of Staff of the Air Force
CYOS	commissioned years of service
DMO	distributed mission operations
FAIP	first assignment instructor pilot
FTU	formal training unit
FYDP	Future Years Defense Program
GAMS	General Algebraic Modeling System
GAO	Government Accountability Office
IFF	Introduction to Fighter Fundamentals
IP	instructor pilot
MAF	mobility air forces
MDS	mission design series
O&M	operations and maintenance
PAA	primary aircraft authorized
PAI	primary aircraft inventory
PMAI	primary mission aircraft inventory
POM	program objective memorandum
RAP	Ready Aircrew Program

RDTM	Rated Distribution and Training Management
RSAP	Rated Staff Allocation Plan
SCM	sorties per crew member per month
SOF	special operations forces
SUPT	specialized undergraduate pilot training
TAMI 21	Transformational Aircrew Management Initiatives for the 21st Century
TARS	total active rated service
TFI	total force integration
TTE	time-to-experience
TWCF	transportation working capital fund
UAS	unmanned aircraft system
UAV	unmanned aerial vehicle
UPT	Undergraduate Pilot Training
UTE	aircraft utilization

Glossary

absorbable unit   a flying unit that accepts inexperienced aircrew members into its crew force

absorption   the process of accessing new undergraduate flying training graduates and/or prior qualified (e.g., first-assignment instructor pilot) aircrews into operational unit line flying positions for their first operational assignments. The Air Force's goal is to balance the long-term need to sustain an inventory that meets requirements against the near-term goal of maintaining unit readiness parameters— that is, to absorb the required number of new aircrews while maintaining at least the minimum unit readiness posture (in terms of experience mix, average time on station, manning levels) required to meet operational taskings and commitments. (AFI 11-412, 2005, para. 3.1.)

absorption capacity   the number of new pilots who can become experienced using the available training resources for a given set of experience and manning policy objectives (which normally would be set by the Air Force leadership)

ALFA tour   short for "ALO, FAC, or AETC tour." A one-time
                     assignment outside a pilot's primary aircraft, after
                     which the pilot returns to the primary aircraft. In
                     the past, these tours have included nonflying posi-
                     tions (such as air liaison and forward air control
                     duties), as well as flying positions as instructor
                     pilots in AETC. ALFA tours can also be served as
                     UAS operators.

air liaison officer an aviator attached to a ground unit who functions
                     as the primary advisor to the ground commander
                     on air operation matters

career enlisted   career field encompassing functions of program for-
aviators   mulation, policy planning, inspection, training and
                     direction, and performing combat operations perti-
                     nent to enlisted primary aircrew activities

distributed   the integration of real, virtual (man-in-the-loop),
mission   and constructive (computer generated) capabilities,
operations   systems, and environments for training. Linking
                     high-fidelity simulators through communication
                     networks so that pilots at different locations can
                     train together is an example of distributed mission
                     operations.

experience	a measure of the amount of time a pilot has in a given aircraft or of the associated skills acquired. For personnel purposes, AFPC uses hours as a metric for experience. For example, a fighter pilot is generally considered *experienced* when he or she has 500 hours of flying time in his or her fighter aircraft. For operational purposes, major commands use the term *experienced* to specify when an aircrew member has upgraded or is ready to upgrade to a flight leadership position (such as aircraft commander, flight lead, instructor). (AFI 11-412, 2005, para. 3.4.6.)
experience level, experience mix	the percentage of a unit's authorized positions that experienced pilots fill (AFI 11-412, 2005, p. 58). The Air Force establishes goals for unit experience levels (for example, 55 percent).
forward air controller	a qualified individual who, from a forward position on the ground or in the air, directs the action of military aircraft engaged in close air support of land forces
line pilots	experienced pilots, with aircrew position indicator 1 (API-1)
overmanning	supplying a unit with more pilots than it is authorized to have based on the number of aircraft that it has been assigned. "[O]vermanning is most often caused by a unit having too many inexperienced aircrew members who need to remain assigned to the squadron to maximize flying opportunities" (AFI 11-412, 2005, para. 3.4.5).

"Pope syndrome"	performance degradation and the loss of combat mission readiness of many pilots due to adverse training conditions. Such conditions, including overmanning of units and low ratios of experienced to inexperienced pilots, existed in A-10 units at Pope AFB in 2000, hence the nickname for the problem. Taylor et al., 2002, describes these conditions in detail.
reserve component	consists of the Air Force Reserve and the Air National Guard
Ready Aircrew Program	annual sortie and event training requirements for fighter and bomber aircrews to maintain combat mission readiness
standard aircraft utilization	also known as *UTE rate* or *standard UTE rate,* measured by number of sorties flown (the average number of sorties flown per assigned aircraft per month) or time flown (the average number of hours flown per assigned aircraft per month). The Air Force leadership establishes standard UTE rate goals. "Comparing the standard UTE rate to actual execution provides Air Force leadership insight into issues impacting real world training such as contingency support or host nation restrictions" (AFI 11-103, 2004, para. 1.1.2).

unmanned aircraft system
what a pilot uses to fly an unmanned aerial vehicle. "That system whose components include the necessary equipment, network, and personnel to control an unmanned aircraft." (JP 1-02) In practice, this is the term the Air Force uses most commonly to refer to the people and things associated with flying UAVs and encompasses the system, aircraft, equipment, and operator, as a whole. By extension, it has come to be the term used for an assignment, and potential career field, piloting unmanned aircraft.

unmanned aerial vehicle
an aircraft that does not carry pilot or passengers. It is a "powered, aerial vehicle that does not carry a human operator, uses aerodynamic forces to provide vehicle lift, can fly autonomously or be piloted remotely, can be expendable or recoverable, and can carry a lethal or nonlethal payload. Ballistic or semiballistic vehicles, cruise missiles, and artillery projectiles are not considered unmanned aerial vehicles." (JP 1-02)

CHAPTER ONE

Introduction

The goal of aircrew management in the Air Force is to maintain an aircrew force whose "size and readiness enable it to accomplish the Air Force mission today and tomorrow" (Air Force Instruction [AFI] 11-412, para. 1.3). This is not an easy task. To accomplish its mission, the Air Force must ensure that it has the right number of pilots in a wide variety of categories, including rank, commissioned years of service (CYOS), types of aircraft flown, and weapon system skills. Training must be provided not only to "absorb" new pilots—that is, turn inexperienced pilots into experienced pilots who can perform a unit's specific combat mission—but also to prepare pilots to acquire the skills required to fill rated supervisory and staff positions at the wing level and above. It can take as long as five and a half years to produce an experienced fighter pilot (U.S. Air Force, 2008). Aircrew management policy changes must therefore be made carefully because their consequences may not be observed for a long time. If the consequences are negative, corrective actions will also need time to take effect. Producing an experienced fighter pilot is also expensive; the cost can exceed $5.7 million.[1]

---

[1] The costs break down to initial pilot training (specialized undergraduate pilot training [SUPT]), $654,062; introduction to fighter fundamentals (IFF), $165,591; the F-15C "basic" course (B-course): $3,453,480. These are projected variable FY 2009 costs through F-15 basic. *Variable costs* are the costs of training additional graduates. These numbers do not include fixed costs, such as military construction, but do include military pay (including student pay), civilian pay, base operating support, temporary duty costs, and travel to the final assignment. (Travel and per diem are derived from factors and are not based on actual data unique to the course.) In addition, is the cost of the fuel required for a new pilot to become

1

For a variety of reasons, aircrew management problems in the fighter pilot community have become acute in recent years. But the primary reason is that the demand for personnel with fighter pilot skills in nonflying jobs, such as positions in air operations centers (AOCs), is increasing while the number of *absorbable* fighter aircraft (aircraft to which inexperienced pilots can be assigned) is decreasing. The need to produce more pilots to meet the demand conflicts with the declining capacity of the system to absorb new pilots and turn them into experienced pilots.

This monograph presents the results of several years of RAND Corporation research that have led to the development of a successful dynamic model of pilot absorption in fighter units. This model has made it possible for the Air Force to assess when and where problems (such as overmanning or unacceptably low monthly sortie rates for inexperienced pilots) are likely to occur in units and to quickly analyze the potential consequences of aircrew management policies designed to solve these problems.

Organization of the Monograph

The next chapter provides a historical perspective on current problems in fighter units, including the complexities of the Air Force system for developing pilots, decisions successive four-star summits have made about pilot-production levels, factors that have made it difficult to recognize the developing problems in fighter units, and problems related to the pilot "bathtub."

Chapter Three describes key issues that affect the mathematical modeling of pilot absorption in fighter units, outlines the development

---

experienced in a unit. This cost could reach $1,513,380 based on the following: An F-15C pilot needs 500 hours of flying time to become "experienced." Approximately 70 hours are flown in the basic course, and 100 hours of simulator time can count toward the total. This leaves 330 flying hours to become experienced. F-15C fuel cost per flying hour when aviation fuel was $2.90/gallon was $4,586. (Headquarters Air Education and Training Command [AETC] FMATT, 2008); AFI 65-503, Table A4-1.)

of the dynamic model used for the analysis in this paper, and presents some examples of model output.

Chapter Four describes how the model was used to analyze the consequences of various aircrew management decisions from 2005 through 2007—some of which had resulted from Air Force Smart Operations for the 21st Century (AFSO-21) initiatives and others from recommendations of the Transformational Aircrew Management Initiatives for the 21st Century (TAMI 21) working group. The key result is that, under policies currently in effect, fighter pilot units will continue to have problems with overmanning, unacceptably low sortie rates for inexperienced pilots, and the ability to turn inexperienced pilots into experienced pilots in their first tour.

Chapter Five shows potential approaches the Air Force could use to satisfy the increased demand for personnel with fighter pilot skills (assuming the increase is justified) and, using the dynamic model, forecasts the resulting improvements in the health of fighter units. Chapter Six presents our conclusions.

Three appendixes provide more background. Appendix A contains the mathematical details of the dynamic model. Appendix B highlights some of the issues raised during the 2005 Aircrew Review of aircrew management issues, and Appendix C describes the recommendations of the TAMI 21 working group.

CHAPTER TWO

How the Crisis in Fighter Aircrew Management Developed

The Aircrew Management Problem

For almost a decade, the Air Force has been undermining the effectiveness of its operational fighter units by overwhelming them with too many newly trained pilots. This is because the Air Force has been trying unsuccessfully over that period to solve another, related problem: a shortage of fighter pilots to fill nonflying rated staff requirements. We will begin this chapter by examining how these issues evolved over time.

The fundamental purpose of aircrew management is to develop and sustain adequate inventories of officers with the operational skills and experience levels needed to meet Air Force requirements. The background the pilots in the inventory have acquired—their years of service, grade levels, weapon system knowledge, mission experience, etc.—should qualify them for the positions they must occupy.

In their effort to fill aircrew requirements for the combat air forces (CAF) following the post–Cold War drawdown,[1] successive Air Force

---

[1] CAF includes fighter, bomber, and other conventional combat resources that underwent substantial reductions in response to diminished threats and resulting budgetary adjustments during the post–Cold War drawdown. Other Air Force resources include the mobility air forces (MAF), consisting mainly of transports and tankers, and special operations forces (SOF), which support and conduct special operations worldwide. We will address MAF and SOF issues as appropriate in this monograph, but the analyses reported here focused primarily on fighter issues because of their relatively large numbers, diminishing force structures, and impending critical problems. Historically, SOF aircrew members have often previously

leadership teams have made aircrew management decisions that seriously degraded the training environments in operational fighter units and even jeopardized their combat capabilities. This monograph documents analyses, beginning in 2005, indicating that the operational fleets of F-15C and F-16 aircraft have been moving toward adverse training conditions that could compromise safety and readiness. Similar conditions—including overmanning of units and low ratios of experienced to inexperienced pilots—existed in A-10 units at Pope Air Force Base (AFB) in 2000 and led to performance degradation and the loss of combat mission ready (CMR) status for many pilots (see Taylor et al., 2002, Ch. Two).[2] The problem stems from Air Force leadership's desire for increasing numbers of new fighter pilots each year; the numbers had reached the point of being greater than the capacity of the operational units to absorb and train them.

Although the problem is simple to state in these terms, the issues remained somewhat obscure to decisionmakers at the time because the absorption capacities of operational units are difficult to assess. Capacities are determined by a number of complex, interrelated factors that were changing fairly significantly over time, while the aircrew management tools then available to the Air Force assumed that the pertinent factors remained constant, reflecting a steady-state environment.[3] These issues provided the motivation for developing models that could accept input values that change with time and that could accurately replicate the system dynamics resulting from changing input values.

---

qualified in another weapon system and moved into the SOF world as experienced crewmembers. Also, until fairly recently, SOF aircrew management relied on MAF resources.

[2] These conditions have become known throughout the CAF as the "Pope syndrome," and we will later introduce descriptive terms to identify the health of operational training environments that include the term *broken* to describe units operating under conditions as bad as those at Pope AFB in summer 2000.

[3] Taylor et al., 2002, identified many of these factors, which we will discuss in more detail later in this monograph. A partial list, however, would include primary mission aircraft inventories (PMAIs), UTEs, unit manning and experience levels, experienced pilot definitions, and flying hour funding issues. Note that the earlier report used the closely related term *primary aircraft authorization* (PAA).

Complexities of the Aircrew Training and Development System

This section broadly examines aircrew management issues and decisions since the end of the Cold War to illustrate the complex, dynamic behavior of the aircrew training and development processes when external circumstances are undergoing rapid, substantive changes. It will also help explain why our model of this process evolved to include certain complexities and dynamic behaviors. Later in this monograph, we will describe the model and the analyses we have performed with its help.

How the Post–Cold War Drawdown Affected the System

The massive drawdown of Air Force forces following the Cold War (Table 2.1) delivered a severe shock to the aircrew management system, and the responses to this shock were responsible for many of the problems we will be discussing.[4]

Facing a surplus of pilots, the Air Force reduced the inventory by lowering production from over 1,500 total active pilots per year in FYs 1989–1990 to about 500 in FYs 1994–1996.[5] There was little concurrent effort to accelerate the departure of pilots already in the inventory.

These reductions in pilot production initially had salutary consequences. Experienced pilots in operational squadrons were relieved of the burden of training new pilots and could concentrate on mastering the most advanced tactics.

By the mid-1990s, however, the fighter pilot inventory had declined far enough that it was necessary to increase production. In addition, a mismatch had inevitably arisen between the supply of and demand for rated officers in specific year groups and grades. Later, we

---

[4] This is not to say that there were no aircrew management issues during the Cold War. However, the drawdown is a convenient starting point for our story.

[5] Pilot production data were provided by Air Force Operations, Plans and Requirements, Operational Training (AF/A3O-AT). A pilot has been *produced* when he graduates from undergraduate pilot training (UPT).

Table 2.1
The Extent of the Post–Cold War Drawdown

Quantity	Average Number for FYs		Change (%)
	1989–1990	1999–2000	
Total active pilots required	22,250	13,603	–39
Active fighter pilots required	7,409	4,747	–36
Active component fighter PMAI	1,959	993	–48
Reserve component fighter PMAI	936	630	–33

SOURCES: Data from AF/A3O-AT; Air Force Manpower, Personnel and Services; Air Force Deputy Chief of Staff for Strategic Plans and Programs, Directorate of Programs; and Air Combat Command, Air and Space Operations, Flight Operations Division.

will discuss certain consequences of this shortage that continue even now to bedevil the aircrew management system.

Four-Star Rated Summit Attempts to Address System Problems

A summit of four-star Air Force leaders, convened late in 1996 to address management problems related to rated pilots, set a steady-state annual pilot-production goal of 1,100 total pilots, of which 370 were to be produced in fighters. These production goals were calculated to sustain inventories of slightly under 14,000 total pilots and about 4,600 fighter pilots, which were the pilot requirements at that time.[6]

A second summit met in April 1999 to address additional aircrew management issues. It had become clear that the total number of fighter aircraft that could accept new fighter pilots (*absorbable aircraft*) was inadequate to absorb 370 new pilots each year without degrading the training environment in the operational units.[7] The summit raised

---

[6] The *sustainment level* is defined as the steady-state inventory generated by the pilot-production goals using historical retention data to determine the expected value for total active rated service (TARS). Taylor et al., 2002, derives the formula and discusses the background more generally.

[7] This aircraft total refers to the primary mission aircraft inventory (PMAI)—the combat-coded airframes that are capable of absorbing new pilots. Examples of nonabsorbable aircraft include the F-117, the F-22 until FY 2008, and aircraft that are coded for training or test

the experience-level objective (that is, the proportion of experienced aircrew position indicator 1 [API-1], or line, pilots) in fighter units to 55 percent (from 50 percent) and lowered the production level for new fighter pilots to 330, which was the new sustainment level based on updated retention and requirements data.[8] It also directed that 30 of the fighter pilots be absorbed in guard and reserve fighter units because 300 was the maximum absorbable number that would ensure the units could maintain at least a 55-percent experience level. The total pilot-production goal remained at 1,100 because other weapon systems did not share the fighter force's structure limitations.

A third summit was convened in June 2001 to address CAF and MAF concerns that the number of new pilots flowing into their operational units was greater than they could properly absorb and develop. Many of these issues, especially for CAF, were attributable to the fact that programs to send new active pilots to guard and reserve units had not been implemented effectively.[9] Indeed, these programs met a great deal of resistance within all three of the total force components for a variety of reasons, primarily associated with funding difficulties and cultural issues.[10] The 2001 summit confirmed the pilot-

---

missions because only experienced fighter pilots can be assigned to fly these aircraft. These numbers are consistent with the analysis documented in Taylor, Moore, and Roll, 2000, drafts of which were available to the summit participants.

[8]   Fighter pilots normally require 500 flying hours in their primary aircraft to become experienced, although there are provisions for pilots with flying experience in other aircraft to qualify with fewer PMAI hours. Inexperienced pilots learn from experienced pilots, so other things being equal, a higher experience level translates to an improved training environment.

[9]   MAF managers had developed a fairly effective program for absorbing new active pilots into guard and reserve flying units, but they were also tasked to absorb the additional 40 new pilots per year that resulted from the reduction from 370 to 330 per year in fighters, coupled with maintaining the annual goal of 1,100 total pilots. Their total force absorption program could not cope with this increase. The corresponding program for CAF units, however, required the new pilots to spend at least one, and normally two, years in an active unit prior to going to a guard or reserve unit, which meant that the bulk of their initial aging process continued to be borne by the active units, and the increased absorption capacity for fighter units was negligible.

[10]   The "total force" comprises the active-duty and reserve components and the Air National Guard.

production goals of 1,100 and 330 for total pilots and fighter pilots, respectively, and established a minimum experience level of 45 percent. This last proviso was added because steady-state analyses showed that, if more than about 300 new pilots entered the system each year (as was the goal), the existing active fighter force structure could not sustain a 50-percent experience level in the operational units, much less the 55-percent target established in the 1999 summit.[11]

Resistance to Realistic Production Limits

Earlier RAND analyses had confirmed that a best-case scenario resulted in a steady-state absorption capacity of 302 new fighter pilots per year, and that under a more realistic constraint, 285 new fighter pilots per year was the limit (Taylor et al., 2002, Ch. Five, esp. pp. 76–81).[12] These numbers were consistent with contemporary analyses conducted by Air Force aircrew managers. Air Force leaders, however, had several compelling reasons to resist recognizing them as acceptable limits.

First, these production levels would yield a steady-state inventory of fighter pilots that would fall some 700 to 800 pilots short of existing requirements. Additionally, coping with a shortfall of that magnitude could require major modifications to the way the Air Force was doing business. It would mean revising the organization and management of its total force components (for example, as we will see later, combining resources from active and reserve components), changing the experience and backgrounds desired for many of its operational assignments, or some combination of these and/or other alternatives.

Second, many senior Air Force officers had been assigned to operational fighter units in the late 1970s and early 1980s. They had personal experience with the successes of the Rated Distribution and Training Management (RDTM) system, which the Air Force introduced

---

[11] These values had been calculated as a part of the analysis documented in Taylor et al., 2002. Taylor, 2004, provides additional details.

[12] These analyses were based on an experience-level objective of 50 percent to ensure that the absolute minimum of 45 percent would not be violated across units. In the more-realistic scenario, aging rates (the rates at which new pilots become experienced) were based on aircraft utilization (UTE) rates that were actually achieved from FYs 1996 through 2000, rather than CAF objective utilization rates.

at that time to cope with aircrew management issues resulting from the post-Vietnam drawdown. These officers consequently believed that analogous, although as-yet-unspecified, changes in training and management processes would enable the fighter fleet to absorb the excess production.

The reduction in operational fighter units after the cessation of hostilities in Southeast Asia, coupled with the lead time required to reduce the production of new aircrew members exiting from an expanded wartime training pipeline, had left the remaining units with excessive numbers of very recently trained (and inexperienced) pilots. The RDTM system addressed experience and manning problems by, first, identifying the absorption problem as the ability of units within a weapon system to accept new pilots and maintain an acceptable experience level. It provided meaningful definitions for what constituted an experienced pilot (fighter pilots are normally considered experienced once they have completed 500 flying hours in their primary aircraft), established acceptable experience-level criteria for operational units (defined as the number of experienced API-1 pilots divided by the total number of API-1 pilots authorized for the unit), and provided quantitative methods for managing training pipelines and assignment processes to ensure that future inventories could match projected requirements. When new airframes, such as the F-15, A-10, and F-16, were introduced in the 1970s, the Air Force was able to maintain higher UTE rates. These higher rates allowed the Air Force to literally "fly its way out" of the post-Vietnam problems.[13]

Thus, many leaders remained convinced that the Air Force could increase UTE and improve management oversight to again fly its way out of the existing difficulties, thereby avoiding having to deal directly with the pilot shortfalls that would result from accepting the absorption constraints.

---

[13] Marken et al., 2007, Chapter Two, develops these issues in more detail. Previously qualified pilots with at least 1,000 flying hours in another Air Force system have an alternative means of becoming experienced fighter pilots, requiring only 300 flying hours in their primary fighter system. Anderegg, 2001, discusses the operational training changes that occurred in Air Force fighter units during the decade after Vietnam far more comprehensively.

This attitude was perhaps accentuated by the prevailing perspective that the current aircrew shortages were primarily the result of the poor aircrew management decisions made early during the post–Cold War drawdown that severely reduced the production of new pilots and navigators. This tended to reinforce the position that the problems could be corrected by reversing the earlier "bad decision" and adopting the "good decision" to produce the "correct" number of new pilots each year, where this correct number was defined by the sustainment levels of 1,100 total pilots and 330 fighter pilots required to produce steady-state inventories that met existing requirements (which in June 2001 had dropped to about 13,350 total pilots and 4,400 fighter pilots, respectively).[14]

Factors Complicating Problem Recognition

Additional factors lulled any sense of urgency about accepting the absorption capacity constraints—and thus the need to mitigate the consequences. The steady-state analyses available at that time could not indicate precisely *when* the absorption crises would actually occur. The analyses merely confirmed that serious problems would indeed exist at some point in the future. Also, a 2001 decision to reestablish standard UTE rates helped slow down the onset of the pending crisis.[15]

Other factors also either delayed the onset of training problems or reduced the leadership's sense of urgency about addressing these issues. Several of these are discussed below.

Pipeline Capacities
In 1996, when the initial four-star rated summit decided to increase undergraduate pilot-production rates, the Air Force had recently com-

---

[14] These numbers are from Air Staff records. Unfortunately, the decrease in fighter pilot requirements was almost totally due to reductions in the primary mission (API-1) billets in operational units that are the core component of the absorption process.

[15] This decision curtailed the steady decrease in home-station UTE rates that had developed after the end of the first Gulf War. See Marken et al., 2007, Figure 2.2, for the changes in UTE rates.

pleted several significant revisions of its UPT programs. The first of these was its reversion to an SUPT program (consisting of separate tracks for pilots headed to different end assignments), followed shortly thereafter by congressionally mandated joint training programs with the Army and Navy to train pilots in helicopters and multiengine turboprop aircraft. The Air Force's total UPT production numbers for the three presummit fiscal years (FYs 1994–1996) were 533, 485, and 523 pilots, respectively. These were the three smallest annual production values the Air Force has ever achieved as a separate service. The new goal of 1,100 pilots per year meant doubling the service's annual UPT production, which was impossible because of another post–Cold War drawdown decision, closing one of the Air Force's UPT training bases. Historically, the Air Force wanted to have a 15 to 20 percent "buffer" for its training capacity, to deal with unforeseen circumstances and changing dynamics, but the base closure left the remaining bases without the capacity to maintain a buffer of any size.[16]

There were similar pipeline issues in formal training units (FTUs), which conduct the B-course training that the active component requires of new fighter pilots before they report to their initial operational assignments. AETC must ensure that every SUPT graduate has a follow-on training slot in an appropriate B-course (or equivalent initial training) program. Constraints in these follow-on programs required careful management of the production numbers. Indeed, as fighter-pilot production numbers began to increase following the 1996 four-star summit, pilots were assigned to mission design series (MDS) aircraft with available FTU capacities, not necessarily those needing new pilots. For example, A-10 production numbers for FYs 1997, 1998, and 1999 were 63, 72, and 80, respectively; the steady-state requirement was only 56 new pilots per year. These disparities were a signifi-

---

[16] See Ausink et al., 2005, Chapter Two (and the references cited there) for more-comprehensive historical information. The first T-1 (as opposed to T-38) trained students completed UPT in July 1993, while the first students completed the Navy's T-44 turboprop training and the Army's helicopter training programs in FY 1995. Post–Cold War UPT production peaked at 1,082 in FY 2002. Data and other information are from AETC/A3R (formerly AETC/DOR). See also Taylor et al., 2002, p. 25, for a discussion of reasons to maintain a buffer.

cant cause of the documented adverse training conditions mentioned above (the Pope syndrome).[17]

The Corona Air Force leadership meeting in fall 2003 examined the effects of pipeline constraints on production goals. The leadership agreed to reduce both total and fighter-pilot production goals by 10 percent, to be implemented incrementally over three years and then restored within the then-current Future Years Defense Program (FYDP). This decision failed to recognize the absorption constraints in the operational fighter units; in fact, after planned cuts in FYs 2004–2006 that would reduce production to 1,000 and 306 in FY 2006, the production goal would build back up gradually to 1,100 and 330 in FY 2009. Figure 2.1 shows Air Force leadership's varying fighter-pilot

Figure 2.1
Primary Mission Aircraft Inventory for Active-Duty and ARC Fighters and Fighter-Pilot Production Goals

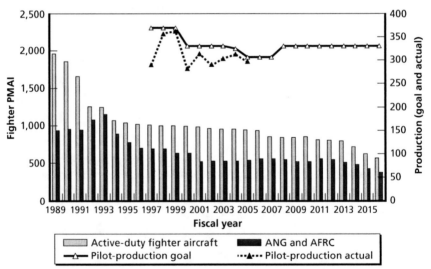

SOURCES: The Air Force Directorate of Programs, Program Integration Division supplied PMAI data for 1989–2001 and Air Combat Command's Air and Space Operations, Flight Management Branch provided the actual production data. PMAI and projected numbers for 2002–2016 are from Bigelow et al., 2003.
RAND MG855-2.1

---

[17] See also the glossary and/or Taylor et al., 2002, Chapter Two.

production goals from FY 1997 through FY 2009 and assumes that the FY 2009 goal continues through FY 2016.

Note the drop from 370 to 330 in FY 2000, the drop to 306 in FY 2006, and the rise to 330 again in FY 2009. The figure also shows *actual* fighter-pilot production (in terms of graduates of IFF) for FYs 1997 through 2005. It is clear that the production goals were never achieved. Finally, Figure 2.1 displays the primary aircraft inventory (PAI) for the active, guard, and reserve components and highlights the fact that the increased FY 2009 production goal would occur when the total number of fighter aircraft available in the active-duty force and the reserve component continued a decline that will, by 2016, reduce it to only 32 percent of what it was in FY 1989.

Ultimately, one or both training pipelines (SUPT and B-course) have continued to constrain new-pilot production to the extent that the production goals set during the rated summits have never been achieved. Indeed, the maximum fighter production value after the 1996 summit was 362 pilots in FY 1999, when the goal was still 370 pilots per year. Total UPT production peaked at 1,082 pilots in FY 2002, even though the 1,100-pilot goal had been in place starting with FY 2000. Fighter-pilot production then dropped off, because of B-course pipeline constraints, to only 288 pilots in FY 2000. It is safe to conclude that the operational training circumstances that were identified at Pope AFB in 2000 would have been more extensive and would have lasted longer had the production goal of 330 fighter pilots per year been maintained. Thus, the full effects of exceeding a realistic absorption limit were not fully realized at that time.[18]

Contingency Support Flying

Another factor that mitigated the effects of exceeding a realistic absorption limit was the Air Force transition into an expeditionary force and the corresponding increase in contingency support flying during the 1990s after the 1991 Gulf War. For example, while the actual number of hours that each API-1 F-15C pilot flew on average per month remained relatively stable during the post–Gulf War period (FYs 1993–1999), the

---

[18] Pilot production data are from AETC/A3R.

number of flying hours that operational units could devote to actual home-station training activities *decreased* by approximately 30 percent. The difference between these values represented the hours that units were required to fly in support of Operation Southern Watch and other contingency deployments. Flying these hours enabled many new fighter pilots to reach the 500-hour experience criterion given in the RDTM system during their initial operational assignments, which they would otherwise have been unable to do if restricted to home-station training sorties. However, unit supervisors recognized that these contingency hours were far less valuable for pilot development than hours flown in normal home-station training operations (contingency flying was often described as "boring holes in the sky").[19]

Contingency tasking for all operational resources increased significantly after the terrorist attacks on September 11, 2001. MAF units, although never as stressed by the new-pilot-production goals as the CAF units, were nevertheless beginning to exhibit problems at about the same time as aircrew managers in the Air Mobility Command (AMC) were preparing for the third four-star rated summit, convened in June 2001. These problems included reduced aging rates (that is, the rate at which new pilots gain operational training in terms of training sorties and hours) and longer intervals before new pilots could become eligible to upgrade to aircraft commander.[20] As the United States prepared to begin combat operations in Afghanistan (and eventually Iraq), flying hours increased markedly to support airlift requirements—flying hours that were paid for through the transportation working capital fund (TWCF). Thus, AMC had a major advantage over Air Combat Command (ACC) because AMC aircraft could increase UTE as required to fly the additional TWCF hours over and above their original flying hour program funding. Fighter units could not enjoy a similar benefit: Because of funding anomalies and UTE constraints, any fighter contingency hours typically must be flown in lieu of normal programmed

---

[19] Marken et al., 2007, Chapter Two develops these issues in more detail.

[20] *Aging rate* is defined as the rate at which new pilots accumulate experience (calculated as the average number of hours per month flown by inexperienced wingmen). See Taylor, Moore, and Roll, 2000.

operations and maintenance (O&M) hours.[21] The effective increase in available flying hours soon after 9/11 allowed AMC to begin increasing its new-pilot inputs. This enabled the Air Force to strive to maintain its total pilot-production goals without egregiously exceeding its fighter pilot training pipeline capacity constraints.

Contingency flying for fighter units increased significantly after 9/11. Initially, the units were tasked to fly defensive combat air patrols during Operation Noble Eagle in support of the new homeland defense initiatives. They were later tasked to begin sequential deployments to conduct combat operations for Operation Enduring Freedom in Afghanistan in 2002 and Operation Iraqi Freedom in 2003. Although the bulk of the Operation Noble Eagle tasking has now reverted to the Air National Guard (ANG), all three contingency-support operations continue as of this writing.[22]

Our model indicates that, were it not for the additional contingency hours the F-15C units flew in FYs 1998 through 2002, they also would have suffered training conditions at least as bad as those that defined the Pope syndrome for the A-10 in FY 2000.

FY 2004 Total Pilot Inventory Match

The annual review of aircrew requirements versus inventory is known within the Air Force as the red-line, blue-line analysis (so called because of the colors conventionally used on line graphs depicting requirements and inventory). It compares existing and projected inventories for each

---

[21] The primary source for information in this paragraph is AMC/A3TF. We should also note that AMC has an internally developed database capability that enables their aircrew managers to track aging rates and upgrade progress continually in each of their units, so they can easily recognize when issues develop. There is no comparable system available for CAF units. Also TWCF hours are funded separately from O&M hours so both can be flown as long as aircraft utilization will permit. Contingency hours for CAF units, however, are never funded until after their O&M hours have been exhausted and then in supplemental legislation that is rarely accomplished in a timely manner. This makes it difficult to fund the majority of these hours and prevents a sizable proportion from being flown even when aircraft utilization rates would permit doing so.

[22] Taylor et al., 2002, and Marken et al., 2007, also address the degraded training per flying hour available to fighter pilots during flights conducted to support operations Southern Watch and Noble Eagle.

aircrew position and weapon system by fiscal year (the blue line) against corresponding projected annual requirements (the red line). Figure 2.2 is an example of a red-line, blue-line display from October 2007. It shows *actual* requirements and inventories through FY 2007, and projected inventories thereafter.

The final issue that mitigated the Air Force's sense of urgency in its approach to aircrew management was the fact that its FY 2004 annual aircrew inventory analysis indicated that, for the first time since the post–Cold War drawdown decisions had been executed, its total pilot inventory numbers matched its total pilot requirements. As seen in Figure 2.2, the inventory actually exceeded the requirements in fiscal years 2004, 2005, and 2006. Unfortunately, the 2004 match of total numbers obscured the significant demographic mismatches within these inventories, especially in terms of weapon system and year of service needs. The data used for the red-line, blue-line analysis can also be used to produce force profiles, such as that in Figure 2.3.

Figure 2.2
Aircrew Requirements Versus Inventory (red-line, blue-line analysis)

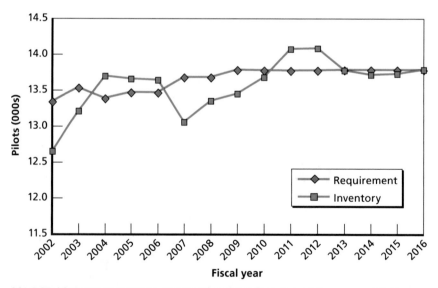

SOURCE: Air Force Manpower, Personnel and Services, Force Management Policy, Military Force Policy, Rated Force Policy (AF/A1PPR), email dated February 25, 2008.
RAND *MG855-2.2*

Figure 2.3
FY 2004 Aggregate Pilot Force Profile by Grade and Commissioned Year of Service

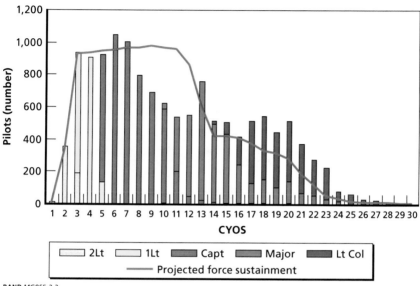

Force profiles compare inventories by CYOS to the projected force sustainment line, which is the CYOS requirements' distribution necessary to yield a sustainable match of inventory to requirements.[23] A gaping hole for officers in the 8 to 12 CYOS groups is evident in the aggregate pilot totals in Figure 2.3. For example, for pilots with nine CYOS, the inventory (about 700) is almost 300 short of the 1,000 required for sustainment. The aggregated data shown do not show the mismatches that exist for different weapon systems. Indeed the fighter

---

[23] This is a subtle point. The red line in Figure 2.3 is the distribution over CYOS that an inventory would achieve if the pilot-production rate remains constant for long enough (nominally 30 years). The Air Force could organize itself so that it required a different CYOS distribution. But as a practical matter the Air Force has defined its billets so they require more-or-less the sustainable CYOS distribution. A deviation of the actual CYOS distribution from the red line, therefore, indicates a deviation from the CYOS distribution of the requirements. One should also recall that aircrew management addresses officers in the grade of O-5 and below so that many pilots are promoted out of the inventory at about 20 CYOS.

pilot inventory remained 200 to 500 pilots short of requirements, depending on how certain categories of requirements were prorated by weapon system, and the CYOS discrepancies were even worse in fighters than for the pilot force as a whole. This meant that the Air Force had overages, for example, in certain cohorts, such as two to five CYOS, consisting mainly of lower-qualified wingmen in the operational units, and 15 CYOS or more, including a substantial proportion of pilots who had not been selected for promotion to the grades of O-5 or O-6. Yet the service still did not have nearly enough fighter pilots in the appropriate year groups to fill either its existing critical staff shortages above wing level or the API-1 instructor and supervisory billets in the operational squadrons.

Despite the demographic mismatches, the total pilot inventory match that finally occurred in FY 2004 received a great deal of publicity, and Air Force leaders may have concluded that the pilot shortage had been resolved and therefore required little additional attention.

Pilot Shortages in Important Year Groups: The Pilot Bathtub

As Figure 2.3 illustrates, the damaged cohorts created by the inadequate undergraduate pilot-production decisions implemented in the 1990s in conjunction with the post–Cold War drawdown stood out so vividly in the charts resulting from these analyses that these cohorts were dubbed the "pilot bathtub" by Air Force leaders and aircrew managers. The demographic mismatches discussed above are largely the result of the persistence of this bathtub as it has shifted each year through the Air Force inventory.[24]

By the late 1990s and early 2000s, the pilot bathtub had begun to adversely affect both the supply of and demand for instructor pilots (IPs). Pilots with eight to ten CYOS, the group that has historically

---

[24] These analyses are managed by the Military Force Policy Division of Air Force Manpower, Personnel, and Services; Air Force Force Management Policy (AF/A1PP); and AF/A3O-AT, which provided the chart in Figure 2.3 and all the information cited in this paragraph.

provided most IPs, were in short supply. The small cohort meant that IPs had to be drawn mostly from other, less-suitable groups. Upgrading pilots with fewer CYOS to IPs has not worked well. But bringing in pilots with more CYOS, who have been out of the cockpit, in staff assignments, also has disadvantages. Moreover, when these pilots were flying, many flew older MDSs and have little experience in the newer fighters.

Moreover, the demand for IPs increased because of the high pilot-production rate. An operational fighter squadron is designed to have a certain number of line pilots (designated API-1) and another number of supervisory and staff pilots (designated API-6). The API-1 pilots are programmed to fly enough to maintain CMR status, while the vast majority of staff API-6 pilots are programmed to fly only enough to maintain the less-well-trained basic mission capable (BMC) status. A unit is supposed to have few enough new pilots that they can all be API-1s with enough API-1 billets remaining for almost all their IPs. Putting too many new pilots into the unit means that more staff API-6 pilots must be IPs, which means they fly more than intended, which in turn means the new pilots fly less than intended (since the unit can generate only a fixed number of flying hours), which means, finally, that it takes longer for the new pilots to accumulate the flying hours they need to be considered experienced and develop the skills required to perform other important functions that require operational expertise.[25]

Once the Air Force allowed the pilot bathtub to develop, it was virtually impossible to overcome its effect. A deficit in pilots with 10 CYOS, for example, cannot be overcome by hiring similar pilots from outside the Air Force; it can only be overcome by ensuring that enough new pilots are trained and retained to fill the gap over time. This dif-

---

[25] Typically, the only non–squadron-assigned API-6 billets that call for CMR, rather than BMC, combat status are the head of the weapons and tactics shop and the chief of the standardization and evaluation division, respectively, and these billets should be filled by IPs. The only other API-6 CMR-designated billets are for squadron-level supervisors (i.e., squadron commanders and squadron operations officers), and the assigned pilots also typically maintain IP status. All other API-6 pilots that maintain IP status fly at monthly sortie rates higher than those normally programmed, taking sorties out of the pool available for training inexperienced pilots.

ficulty was exacerbated by taking fighter force structure cuts disproportionately from active rather than reserve resources. The decision to do so may have been based largely on the political realities surrounding where the Air Force could actually take the required cuts in bases, units, and force structure, but the results ensured that the ensuing problems could not be resolved using conventional means. No surge capacities remained that would permit corrective action, especially when, as we saw in Figure 2.1, the actual PMAI numbers for the active and air reserve components were declining.

Conclusion

Continued efforts to correct the bathtub and its consequences caused the Air Force to endeavor to operate its new-pilot-production system at or above training pipeline capacities or absorption capacities, or both, for the better part of a decade. The resulting issues finally led to a three-star review of aircrew management, chaired by the Deputy Chief of Staff for Operations, in December 2005. Our analyses, presented at that review, showed the consequences of then-existing aircrew management policies. Before providing detailed examples of the analyses, a description of how our aircrew management model works is in order.

CHAPTER THREE

Modeling the System

RAND has been modeling the absorption of pilots in the Air Force for several years. This chapter provides a brief overview of the dynamic model we used to produce the results we presented at the 2005 Aircrew Review and later to analyze policies proposed to solve the problems highlighted at that meeting. Appendix A presents the technical details of the model.

A Steady-State Picture of Pilot Absorption

Prior to FY 2005, all models used to assist aircrew managers considered the inventory of rated officers to be a steady state (hereafter, we will consider only rated officers who are fighter pilots). By this, we mean that the size of the inventory remains constant over time, and new pilots are added to the inventory at a constant rate. These two assumptions imply that pilots leave the inventory at the same rate that new pilots are added. For purposes of our model, we defined the following variables:

$Rqmt$ = number of fighter pilots required

Inv = inventory of fighter pilots

$Prod$ = new fighter pilots added per year (production rate)

$TARS$ = average years a fighter pilot remains in the inventory.

When the inventory is in steady state,

$$Inv = Prod \times TARS. \tag{3.1}$$

As mentioned in Chapter Two, the purpose of aircrew management is to develop and sustain adequate inventories of officers with the requisite operational skills and experiences needed to meet Air Force requirements. One aspect of this, the one on which the four-star rated summits focused the greatest attention, consists of making the inventory equal (or as nearly equal as possible) to the requirement, or

$$Inv = Rqmt. \tag{3.2}$$

Aircrew managers have greater influence over *Prod* than over either *Rqmt* or *TARS*, so most of the time they will take *Rqmt* and *TARS* as given and adjust *Prod*. Combining Equations 3.1 and 3.2, shows that matching the inventory to the requirement requires that

$$Prod = \frac{Rqmt}{TARS}. \tag{3.3}$$

A number of factors constrain the production of fighter pilots. To become a fighter pilot, an Air Force officer must first complete SUPT, where he or she learns basic flying skills. After some intervening activities,[1] he or she attends FTU, where he or she takes the B-course to learn how to fly one of four different fighter aircraft: the A-10, F-15C, F-15E, or F-16.[2] Graduation from the B-course is synonymous with production.

On graduation from the B-course, the pilot is assigned to an operational fighter squadron. A pilot is absorbed into the fighter pilot inventory during this assignment. To be absorbed, a pilot must acquire the skills and experience needed to perform well in subsequent assignments, both flying and nonflying. Books might be written about just what these skills and experiences should be, but for management purposes, the Air Force assumes that a pilot with 500 flying hours in the

---

[1] For most pilots, the intervening activities take six months or less. Some pilots, however, serve a three-year tour as instructors in training aircraft before they attend the B-course. These are referred to as *first-assignment IPs* (FAIPs).

[2] They will soon be joined by a fifth fighter, the F-22. UPT graduates were first assigned to F-22s in 2008; the first group completed F-22 training in November 2008. Until that first group graduated, only experienced fighter pilots were being assigned to the F-22.

fighter will have acquired them.[3] A pilot with less than 500 flying hours is deemed inexperienced; after that, he or she is deemed experienced.

Our models have focused on the third step, the capacity of the operational fighter squadrons to absorb new pilots. In the steady-state picture, the number of flying hours inexperienced pilots need to meet the criterion to become experienced cannot exceed the number of hours available to them, and this constraint is expressed as follows:

$$FH2E\left(UPT\right) \times Prod\left(UPT\right) + FH2E\left(FAIP\right) \times Prod\left(FAIP\right) \leq FHinex, \quad (3.4)$$

where $FH2E$ (UPT) and $FH2E$ ($FAIP$) are the number of flying hours a UPT graduate (i.e., a pilot who enters the B-course immediately after graduating from SUPT) or a FAIP respectively, must accumulate in the operational squadron to become experienced; $Prod$ (UPT) and $Prod$ ($FAIP$) are numbers of UPT graduates and FAIPs produced per year; and $FHinex$ is the number of flying hours per year available for inexperienced pilots.[4] The production rates are the quantities constrained by this inequality.

Equation 3.4 is the *absorption constraint.* If production levels remain constant, policies for relaxing this constraint must focus on either increasing the flying hours available to inexperienced pilots or reducing the number of flying hours a pilot requires to become experienced.

Returning to Equations 3.1 through 3.3, the Air Force itself can influence both *Rqmt* and *TARS* , although perhaps not so much as the people primarily charged with aircrew management can. The number of fighter pilots required is simply the number of billets that the Air Force allocates to fighter pilots, inflated to account for the inevitable fraction of pilots that are between jobs or in school. But many of those are nonflying billets, for which experience as a fighter pilot might be

---

[3] This includes the 70 to 80 flying hours accumulated during the B-course. For a FAIP, the figure is 300 flying hours in the fighter.

[4] The aircraft assigned to an operational fighter squadron fly a limited number of hours, less than half of which can be flown by inexperienced pilots. For the most part, this reflects the fact that most inexperienced pilots are qualified to fly only as wingmen. Almost all sorties by a wingman must be accompanied by a flight-lead–qualified pilot flying another aircraft.

helpful in them but may not be absolutely necessary. Thus billets can be (and have been) moved into or out of the requirement for fighter pilots based on expert judgment.

The usual means for influencing *TARS* is to adjust the bonus fighter pilots receive for remaining on active duty. If the bonus is increased (decreased), it is expected that more (fewer) pilots will remain. The bonus can even be negative—paying pilots to separate rather than remain. Factors beyond the control of the Air Force (especially job opportunities with commercial airlines) also affect whether fighter pilots choose to separate, so using the bonus to manage *TARS* is imprecise.

Two other, extraordinary, methods have been used to influence *TARS*: stop-loss orders and changing the active-duty service commitment. Following the attacks of 9/11, the Air Force issued a stop-loss order, which prevented fighter pilots from separating from the Air Force for two years.[5] Becoming a fighter pilot entails a commitment to a specific length of active duty. This commitment can be changed and has been, from 6 to 8 years in the mid-1990s and from 8 to 10 years in 2002 and 2003. Neither of these extraordinary measures can be taken very often.

The Dynamic Picture of Pilot Absorption

By 2005, we had exhausted the potential of steady-state models to analyze aircrew-management problems. Since the post–Cold War drawdown in the early 1990s, aircrew managers have been forced to cope with rapid reductions in requirements for fighter pilots and in the number of fighter aircraft in the force structure. Steady-state analysis is not adequate to portray the consequences of these changes or of the responses to them, all of which have ensured that the fighter pilot inventory is far from steady state, both in overall numbers and in distri-

---

[5]   The Air Force ended its stop-loss program in June 2003. The career fields subject to stop-loss changed several times between January 2002 and June 2003, but selected pilots were affected for the entire period. See Government Accountability Office (GAO), 2004, p. 76.

bution over years of service, grade levels, and other markers of experience (e.g., the pilot bathtub discussed in Chapter Two).

For both the steady-state and dynamic pictures, we built separate models for each absorbable aircraft. For each steady-state model, we could represent the inventory of fighter pilots by a single number. For each dynamic model, we needed to represent the inventory of pilots in each period (for our purposes, a month); to track the flow of pilots from one month to the next, we needed to distinguish pilots by the number of months since they had graduated from the B-course. We tracked pilots that had entered the B-course directly out of UPT separately from FAIPs, who had entered the B-course after an intervening tour as IPs in UPT.

Initially we deemed it sufficient to track pilots only through their first operational tour, which could last up to 36 months. We simulated the FY 2000 through FY 2025 inventory (a total of 26 years, or 312 months); tracking pilots through their first operational tours for this span of time required calculating 22,464 ($312 \times 36 \times 2$) different segments of the fighter pilot inventory for each MDS. The model calculated the number of other pilots—pilots in their second or a subsequent operational tour—needed to man the operational squadrons according to the policies of the aircrew management community and simply assumed that these pilots would be available.

In later versions of the model, we also tracked pilots through their second operational tours. For second-tour pilots, we dropped the distinction between UPT graduates and FAIPs, so tracking second-tour pilots meant calculating another 11,232 (312×36) segments of the inventory.

Thus, each dynamic model is much larger than all the steady-state models combined. Indeed, we implemented all the steady-state models in a single Excel spreadsheet.[6] We implemented each dynamic model as

---

[6] Bigelow et al., 2003, documents the steady-state models and shows how the small "repro" model we implemented in an Excel spreadsheet can closely approximate a detailed linear programming model of operational training.

a deterministic simulation in the General Algebraic Modeling System (GAMS).[7]

The model steps through time in intervals of one month. In each month, it calculates the number of first-tour pilots in the inventory starting from the previous month's inventory and the UPT and FAIP production rates for the current month (assumed to be 1/12th the annual production). It adds second- and subsequent-tour pilots to the operational units based on authorizations for API-1 and API-6 pilots and the target experience level. It calculates the flying hours available during the month from the PMAI, the UTE rate, and the average sortie duration. Then it iteratively adjusts the number of hours each inexperienced pilot flies to bring the demand for inexperienced flying hours into balance with the supply in that month.

In the steady-state picture, the number of pilots that can be absorbed annually is limited only by Equation 3.4. Absorption must equal production in steady-state, so the UPT and FAIP production rates are also constrained by Equation 3.4. In the dynamic picture, instead of a single constraint, there is a "web" of constraints. Each month has its own constraint on flying hours available to inexperienced pilots. Inexperienced pilots fly in a succession of months until they have accumulated the requisite number of hours to be deemed experienced. It is therefore possible to overproduce pilots in some months and to make up for that by underproducing them in other months. Likewise, it is possible for flying hours to decrease in some months so long as they increase in other months. Despite this flexibility, only a given number of flying hours is available over the long term, and attempting to produce too many pilots will, as represented in the model, cause a variety of problems.

---

[7]   If *deterministic simulation* sounds like an oxymoron, it is because of the common assumption that a simulation model must have random elements, a type often called a Monte Carlo simulation model. Our model is a simulation with no random elements, which is why we call it a deterministic simulation.

GAMS is generally used to generate the large data structures used in mathematical programming (i.e., optimization) models, and the GAMS application is distributed with a number of powerful solvers for such problems. Our model makes use of GAMS' array manipulation capabilities, but involves no optimization at all.

An Example of the Capabilities of the Dynamic Model

Some of the cases we generated for F-15C pilots during an early analytic exercise to examine the consequences of different levels of production will illustrate the model's capabilities. Table 3.1 shows the pilot-
production rates for three cases. The base case used the production
rates in the SUPT distribution plan we received from AF/A3A-OT
(the AF headquarters office of Aircrew Management) in May 2005. For
the red case, we reduced the numbers of pilots entering the B-course
directly from SUPT (the UPT graduates) enough to ensure that they
became experienced before they were a full 36 months into their first
operational tours. For the blue case, we further reduced UPT graduates
to ensure that the manning levels of operational units never exceeded
110 percent of authorized manning. The number of FAIPs entering the
B-course was the same for all cases.

Figures 3.1 through 3.4 show the four model outputs that we considered most important for these cases. Figure 3.1 represents the manning of operational squadrons as a percentage of authorized manning.
In the base case, unit manning exceeds 120 percent by FY 2008; the
reductions in pilot production for both the red and blue cases ensure
that unit manning remains below 110 percent from FY 2007 through
FY 2011. Figure 3.2 represents the time to experience for UPT gradu-

Table 3.1
Pilot Production for Three F-15C Cases

| | UPT Graduates | | | FAIPs |
FY	Base Case	Red Case	Blue Case	All Cases
2004	56	56	56	19
2005	47	47	47	18
2006	53	53	46	20
2007	45	33	33	16
2008	35	20	20	12
2009	44	36	6	4
2010	38	30	30	4
2011	32	30	28	4

Figure 3.1
Example Model Output: Manning of Operational Squadrons

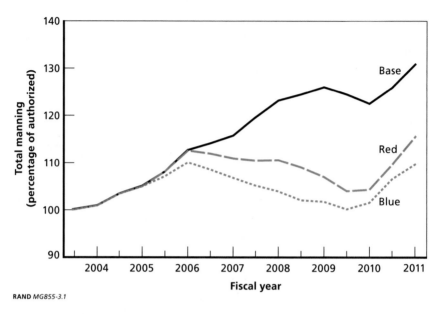

Figure 3.2
Example Model Output: Time to Experience for UPT Graduates

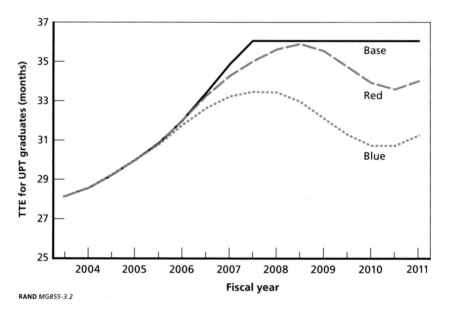

Figure 3.3
Example Model Output: Sorties Each Inexperienced Pilot Flies per Month

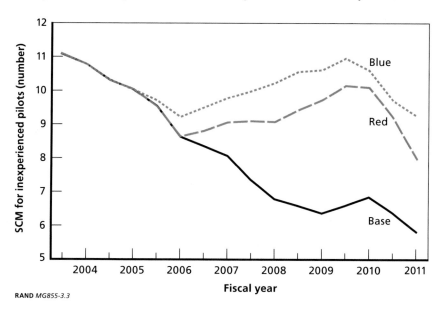

Figure 3.4
Example Model Output: "Real" Experience Level

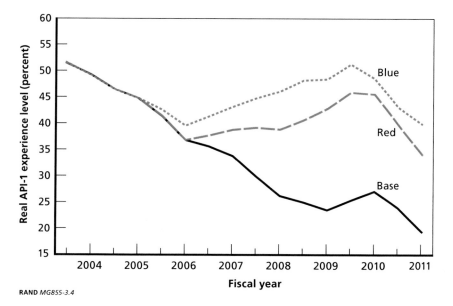

ates. Here, the base case exceeds 36 months by FY 2008,[8] while the red and blue cases ensure that pilots can become experienced within a three-year tour. Figure 3.3 shows the sorties each inexperienced pilot flies per month. Bigelow et al. (2003) showed that the desired number is 12 or 13 sorties per month, with ten sorties per inexperienced pilot per month a bare minimum. The red case is an improvement over the base case, but only the blue case ensures that inexperienced pilots can fly the bare minimum for an extended period. Figure 3.4 shows the "real" experience level. This is the ratio of experienced to total pilots occupying API-1 slots. It is different from the "book" experience level (the level defined in AFI 11-412), which is the ratio of experienced pilots in API-1 slots to authorized API-1 slots. The book experience level exceeds the real level whenever squadrons are overmanned and thus gives an unrealistically rosy picture of the units' health. These figures show that none of the cases achieve the goal of a 55-percent experience level.

All versions of the model can generate many other outputs, including

- the number of pilots not fully experienced by the end of their first operational tours, by year
- the percentage of API-1 billets occupied by first-tour pilots; a high percentage (over about 80 percent) suggests that the unit will have difficulty providing experienced pilots during deployments.

Versions of the model that track second-tour pilots can generate additional outputs, including

- the number of pilots awaiting a billet for a second operational tour or the equivalent
- the time a pilot will spend waiting for a second-tour billet, as a function of the time the pilot first became eligible for a second tour

---

[8] The model does not allow pilots to remain in a unit after 36 months, so the horizontal line at 36 months starting in FY 2008 for the base case means that pilots leave the unit after 36 months without becoming experienced.

- the time a pilot did spend waiting for a second-tour billet, as a function of the time the pilot was assigned to a second-tour billet or the equivalent.

Evolution of the Dynamic Model

We have used the model in a number of analytic exercises since we first implemented it. The model has required modifications for each successive exercise to address different possible aircrew management policies. The remainder of the section describes the various versions of the model and the policies they were designed to assess.

Varying Pilot-Production Rates

Adjusting the pilot-production rates manually for the red and blue cases shown in Table 3.1 and Figures 3.1 through 3.4 was laborious. So, we automated the procedure in early 2006. This allowed us to set a target value for unit manning as a percentage of authorized manning, and the model would automatically seek month-by-month production rates that would achieve the target. We also experimented with setting targets for other quantities, such as the time it took for a new pilot to become experienced.

The model performed as designed—it met the specified target if it was allowed to select a production rate from a wide enough range—but the results were nonetheless unsatisfactory. Figure 3.5 shows that, for the F-15C, the model was able to achieve target levels of 100, 105, and 110 percent of authorized manning starting in FY 2007. However, the production rates required to achieve even a constant manning target would typically fluctuate wildly, as seen in Figure 3.6. The problem arose because the model tries to correct a shortage from the target by adding as many pilots as necessary, up to the maximum allowed, or tries to correct an overage by adding as few pilots as necessary, down to the minimum allowed. Thirty-two months later, this large (or small)

Figure 3.5
Manning Levels in Cases with Target Manning

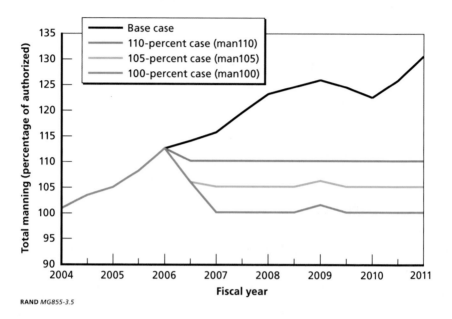

cohort exits the unit en masse, and the model must compensate by adding another large (or small) cohort of new pilots.[9]

Large fluctuations do sometimes occur in the real-life number of pilots entering or exiting a squadron, and they disrupt unit operations. We therefore experimented with various ways to smooth the production rates, but we have yet to devise a satisfactory procedure. Accordingly, we have used automated calculation of pilot-production rates sparingly.

---

[9] The small increases in Figure 3.5 in FY 2009 for the 100 and 105 target levels arise from the fact that the model forces every pilot to complete his 32-month tour before he leaves the unit. If the force structure is declining, the *requirement* for that pilot may disappear (because there are fewer aircraft) before he completes his tour. Since the pilot remains, the percentage of personnel compared to authorized personnel increases.

Figure 3.6
Entry Rates Needed to Achieve Target Manning

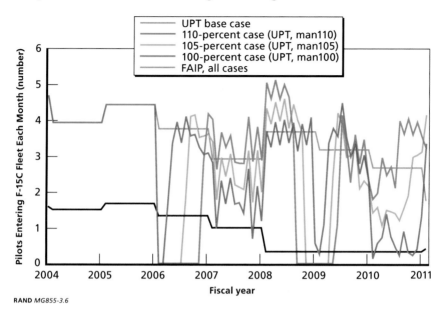

RAND *MG855-3.6*

Allowing Simulator Time to Count Toward Experience: Model Updates in 2006

During the first half of 2006, we extended the model so it could assess the effect of allowing pilots to count sorties "flown" in a simulator toward experience hours. We implemented two approaches. One was to reduce the number of hours a pilot would have to fly in an actual aircraft to become experienced, for example from 500 to 400. To implement this approach, we allowed this number to vary over time. (In the initial model, the number of flying hours required to become experienced is a constant.) The reduction would occur at the time the simulator credit is phased in.

The second approach was to increase the number of flying hours available, as if time spent in the simulator can be considered the equivalent of time spent flying the actual aircraft.[10] The model calculates the

---

[10] It is not necessary to assume that an hour of simulator time is the equivalent of an hour of flying time. One could, for example, increase the flying hours available by one hour for every

number of flying hours available as the number of PMAI aircraft multiplied by the sorties per aircraft per month (the UTE) and the average sortie duration (ASD). In the initial version of the model, the UTE and ASD were constants. To implement this approach we allowed the UTE to vary over time.

Conversations with Air Force fighter pilots suggested, however, that they were skeptical that 400 actual flying hours plus the equivalent of another 100 flying hours in a simulator would truly give a pilot the same level of skills and experience as 500 actual flying hours. We cannot answer that question, but we did change the model to explore some possible consequences of this skepticism.

In the initial model, the hours required to become experienced influenced three aspects of a pilot's first operational tour:

1. As explained earlier, pilots who are inexperienced have access to less than one-half the hours that all pilots in the unit fly; if there are too many inexperienced pilots, each will fly less than the pro rata share.
2. The policy for manning the unit specifies a minimum number of experienced pilots that must be assigned to the unit; if there are too many inexperienced pilots, the actual manning will exceed the authorized manning.
3. A pilot's first operational tour is planned to last 32 months; if a pilot has not become experienced by the end of that time, his tour can be extended.

We changed the model so we could use two different definitions of experience. To represent the skepticism that a pilot is truly experienced without a full 500 hours of actual flying, we can use the old 500-hour definition to determine how many pilots have restricted access to flying hours. But we can use the new 400-hour definition for the other two calculations, i.e., for manning and whether or not a tour needs to be extended.

---

two hours of time available in the simulator.

Another Model Update in 2006: Incorporating Second Operational Tours

In April 2006, an email was forwarded to us from the commander of an Air Force fighter squadron. It began: "[My squadron] is very short on IPs."

IPs are essential to a fighter squadron's operation. A pilot qualified as a flight lead can supervise wingmen during ordinary training flights, but only an IP can supervise upgrade sorties. Without an IP, the wingman cannot upgrade to a flight lead, and a flight lead cannot upgrade to an IP.

We have long been concerned that maintaining the pilot-production rate high relative to the number of PMAI aircraft has adverse consequences on more than absorption. In particular, the number of fighter pilots per aircraft will rise, leaving the average pilot fewer total flying hours over his entire career. This might lead to a shortage of highly experienced pilots, those with 1,000 flying hours or more. We wondered if the shortage of IPs noted by the squadron commander was a symptom of this potential problem.

Few pilots become IPs until they are well into their second operational tours. To investigate this question, therefore, we extended the model to track pilots through their second operational tours. A pilot leaving his first tour is placed, after a delay,[11] in a pool of pilots eligible for a second assignment to an operational squadron. The model assigns these pilots to operational squadrons as needed, and once assigned, a second-tour pilot remains in the unit for 32 months.

Cases run with the new model suggested that there is no reason to expect a persistent shortage of IPs, although there can be episodic shortages. These can arise if the rate at which second-tour pilots are assigned to operational units fluctuates widely over time, as it will do if the rate at which first-tour pilots are assigned to operational units fluctuates widely over time. Historically, about one-half the second-tour pilots eventually become IPs, and this occurs several months after the

---

[11] For most pilots the delay is three years, which represents the time spent in an intervening assignment, such as serving as an air liaison officer (ALO) attached to an Army unit. Only about 10 percent of pilots are assigned a second operational tour immediately after their first tour; for them, we introduce a delay of six months.

start of the second tour. If a large cohort of second-tour pilots exits the unit, they will be replaced by an equally large cohort of new second-tour pilots, but for the first several months, none of these will be serving as new IPs to replace the departed IPs. This episodic problem could be addressed by smoothing the entry rate of second-tour pilots or by managing the lengths of pilots' second tours so as to smooth the departure rates of IPs. At present, neither of these options is implemented in the model.

The cases did suggest that there is a long-term problem looming for pilots eligible for a second tour, however. As the number of PMAI aircraft declines, the number of billets available for second-tour eligibles also declines. It may be that not enough billets will be available to ensure that all who are eligible for second tours will get one.

To investigate this issue, we modified the model further. We added a provision for substitute second-tour billets, billets (e.g., FTU instructor) to which a pilot might be assigned in lieu of a second tour in an operational squadron. We also added calculations of (1) the time a pilot just assigned to a second tour (or a substitute) has waited in the pool of eligibles and (2) the time a pilot just entering the pool of eligibles will wait for such an assignment.

New Model Capabilities in 2007: Accounting for Early Departures

In early 2007, yet another idea for restoring the operational squadrons to health without reducing the fighter pilot-production rate emerged. This one called for some first-tour pilots leaving their units early, before they had accumulated 500 flying hours. One possible destination for these individuals would be piloting unmanned aerial vehicles (UAVs) as a career, never to return to fighters. A later variant of the idea called for some second-tour pilots to leave the unit early as well, before their 32-month tours were completed.

We therefore added the ability to specify the number and timing of early departures of both first- and second-tour pilots to the model. The user specifies the number of pilots to depart early in each month of the simulation and a window of time on station during which they will depart.

The model results can shed light on the benefits of early departures for the pilots left behind, such as higher numbers of sorties flown per pilot per month and a shorter TTE. The model makes no attempt, however, to estimate the consequences for the pilots who depart early. It does not track these pilots to whatever assignment they receive on departure. Neither does it calculate when, if ever, the pilots receive a later assignment flying a fighter aircraft. Rather, the departing pilots drop out of the model entirely.

The Forever-Unfinished Model

We have no current plans to make further changes, but considering the model's history, we cannot rule them out. It appears that this model, like many others, will reach a final configuration only when it outlives its usefulness.

CHAPTER FOUR

Air Force Policy Decisions: 2006–2008

As Chapter Two explained, the Air Force has been facing a dilemma for almost two decades: either undermine the capabilities of its operational fighter units by overwhelming them with too many newly trained pilots or accept a shortage of fighter pilots to fill certain nonflying but nevertheless critical positions. In this chapter, we discuss the attempts the Air Force made from 2006 through 2008 to resolve this dilemma. Our dynamic model supported the Air Force policy deliberations at many points, providing estimates of the effects that proposed policies would have on the health of operational fighter squadrons.[1]

To maintain a healthy operational and training environment in the units, the flow of new pilots must be commensurate with the units' capacities. One approach is to reduce the flow of new pilots. Over time, this will reduce the number of pilots in the overall inventory, and thus exacerbate the shortage of pilots to fill nonflying positions. Air Force leadership has resisted this solution, acceding only to temporary reductions or temporary diversions of inexperienced pilots from the units.

One alternative is to increase the capacity of the system to accommodate new pilots. A number of policies dealt with in this chapter have this effect. These include giving flying credit for simulated sorties, assigning some active pilots to ANG units to gain flying experience,

---

[1] This is only one side of the equation, of course. The other side consists of the policies' effects on the shortage of fighter pilots in nonflying positions. It is fairly simple to estimate the policies' effects on the size of the shortage, but it is very difficult to measure more meaningful effects of the shortage (e.g., on the quality of Air Staff and major command staff decisions).

and making unmanned aircraft system (UAS) billets absorbable (i.e., capable of taking new pilots directly out of UPT and turning them into experienced pilots).

Even as aircrew managers were attempting to balance the need for more fighter pilots against the health of the operational units, other factors were making the problem harder. For example, cuts in the O&M budget for flying hours and reductions in the projected fighter force structure reduced the capacity of the system to absorb new pilots.

We will now describe in detail how these themes played out from FY 2006 through FY 2008 and the analytic support we provided the process.

Aircrew Review 2005

At Aircrew Review 2005, held in December 2005, we presented an analysis, summarized here, of the options as we saw them (see also Appendix B).

Characterizing the Health of Fighter Units

We used the following terms to describe the health of operational fighter units:

- A *healthy unit*
  - has a manning level of 100 percent, to remain manned very close to the total number of authorized (API-1 and API-6) pilots
  - has an experience level of 60 percent; in essence, all the assigned API-6 and about 60 percent of the assigned API-1 pilots are rated as experienced, enabling inexperienced pilots to fly at the same rate as the unit's average for CMR pilots.[2]
- A *stressed unit*
  - is overmanned, but the manning level is monitored and controlled at between 105 and 110 percent

---

[2] Here, *experienced* is defined in terms of the appropriate AFI 11-2, *Flying Operations*, Vol. 1 for the aircraft.

- has an experience level of about 45 percent; only 45 percent of the API-1 pilots are experienced, restricting the sorties available to inexperienced pilots to the extent that they struggle to maintain minimum CMR requirements.
- A *broken unit*
  - has a manning level exceeding 120 percent
  - has an experience level below 40 percent; many new pilots are not able to fly often enough to become experienced in a 36-month tour.

Our earlier models indicated that new pilots cannot fly sufficient sorties per crew member per month (SCM) unless they are in units that are manned at (or very near) their authorized levels and that possess adequate numbers of experienced pilots to provide the required in-flight supervision. Our examination of manning and qualification data available from the operational units for FYs 1998–2000 and our discussions with assigned instructors and supervisors indicate that this experience requirement matched an actual experience level of 60 percent (i.e., that 60 percent of the assigned API-1 pilots had to be experienced). A healthy unit, one with these characteristics, can distribute training sortie resources fairly uniformly among all CMR pilots (Taylor, Moore, et al., 2000, and Bigelow et al., 2003).

Similarly, stressed units may be overmanned, but the assignment process monitors and controls manning levels to ensure that they do not exceed 110 percent. These conditions reduce flying rates for inexperienced pilots because experienced pilots must fly more often than their own programmed needs would dictate because they provide the essential in-flight supervision for the less-experienced pilots. This stresses the units as experience levels continue to decrease. The resulting inertia in the system and the time required to adjust outputs from the training pipelines mean that assignment actions must start early (at about 105 percent manning) to avoid breaching the 110 percent limit. The Air Force requires the assignment process to maintain units at *reported* (or "book") experience levels of 50 percent, but these rates are measured relative to the *authorized* number of pilots (AFI 11-412, 2005). With overmanning, the actual number of pilots exceeds the number authorized. This means that the *actual* experience levels (measured as the percentage of assigned API-1 pilots) will often drop to 45 percent or less as manning levels reach the 105 to 110 percent range, even if the "book" experience level remains at 50 percent or more. It is interesting to note that many active fighter units have continued to operate under stressed conditions for the vast majority of the period since the aircrew summit in 1996.

The unit at Pope AFB that we mentioned in Chapter Two was *broken* in July 2000. This term typically describes indications that the aging rates of new pilots have decreased so far that they can no longer achieve the experience criterion (per AFI 11-2, Vol. 1) for their primary aircraft within the initial operational flying period, which is normally limited to 36 months. Both our model runs and our observations at Pope AFB indicate that manning levels typically exceed 120 percent and actual experience levels typically drop into the 30 to 40 percent range as this condition occurs. It is important to note that these conditions continue to worsen unless (and until) specific corrective action is taken. Another important observation is that, by the time an entire system breaks, its units will have been operating under extremely stressed conditions for an extended period. No weapon system was broken fleetwide when Pope AFB experienced its difficulties in 2000. Corrective initiatives had already been implemented at Pope AFB, and conditions were beginning to improve there at the time of our site visit. A broken weapon system, as opposed to a broken individual squadron, implies broader, deeper, and longer-lasting difficulties than those that occurred in 2000. We will address some of the implications of these kinds of conditions later.

Model Results Presented to Aircrew Review 2005

The slowdown in flying rates for new pilots as unit manning increases and unit experience decreases is a major contributor to the system dynamics governing the flow of new pilots through the operational units. Air Force absorption and assignment models have never addressed the adverse effects of this dynamic behavior. This was a primary consideration that motivated us to develop dynamic models that could capture the implications of these changing dynamics effectively.

We ran our models for every MDS that was capable of absorbing new pilots or would become capable of doing so during our period of interest. Thus, we developed F-22 and F-35 models, even though only the A/OA-10, F-15C, F-15E, and F-16 were actually absorbing new pilots at the time. These straightforward cohort models move pilots month by month through operational units in a manner governed by a fixed set of constraints and parameter values. We tuned our initial conditions to reproduce actual data reasonably closely; where assumptions were required for parameter values, we attempted to be conservative, in the sense that the resulting training environment would be at least as bad in the real world as the model runs indicate for the same conditions.

Table 4.1 summarizes the results of these model runs. F-15Cs were projected to break by the end of FY 2007 and the F-16s would break during FY 2009. The model results were based on parameter values that were sufficiently conservative, however, that we concluded that they represented "not later than" dates by which these systems would be broken. Indeed, our analysis indicated that they could potentially break several months prior to the dates indicated in the table. Such was the situation before two decisions in 2006 that affected the health of fighter units.

Table 4.1
Summary of Model Runs for Absorbable Fighters in December 2005

	FY 2004	FY 2005	FY 2006	FY 2007	FY 2008	FY 2009	FY 2010	FY 2011
A/OA-10	Str	Str / Hlth	Hlth	Hlth	Hlth	Hlth	Str	Str
F-15C	Hlth	Str	Str	Str	Brk	Brk	Brk	Brk
F-15E	Hlth	Hlth	Hlth	Hlth	Str	Str	Str	Str
F-16	Str	Str	Str	Str	Str	Str / Brk	Brk	Brk

NOTE: Crew ratios for the OA-10 decreased from 2.0 to 1.5 in FY 2005, and the production of new A/OA-10 pilots consequently decreased as well. This adjustment benefited these units while increasing the flow of new pilots into F-15C and F-16 units, further degrading their training.

Healthy	Hlth
Stressed	Str
Broken	Brk

The Effects of Crediting Simulator Time and Related AFSO-21 Policy Decisions in 2006

In early 2006, the Air Force established its AFSO-21 program, which examined the idea of allowing certain activities performed in advanced simulator systems to substitute for flying activities in meeting pilot experience criteria. Hours logged in a simulator accomplishing Ready Aircrew Program (RAP) tasking-memorandum-approved missions could be counted as flying hours as long as they met specified conditions and did not exceed 20 percent of the total number of hours required to meet the threshold for experience (AFI 11-2F-15, p. 71). Thus, a pilot with 400 aircraft hours and 100 simulator hours would satisfy the requirement for 500 hours in the primary aircraft. At the time, the F-15C was the only fighter with fully outfitted mission-training centers readily available within its active operational units, so this was a way to formally recognize the value of new high-fidelity simulators and distributed mission operations (DMO).[3] In addition to recognizing the importance of advanced simulator training, it was hoped that this decision would also make it easier for inexperienced pilots to become experienced in their initial operational flying tours.[4]

The aircrew and training managers who developed this proposal hoped that it might alleviate the effect of the absorption-capacity constraint in the operational units. They planned to continue producing new F-15C pilots at the programmed rates because they felt the new experience definition would enable these new pilots to become experienced more quickly and prevent the units from being overwhelmed with inexperienced pilots. Recall that these units were projected to break in FY 2007 for precisely this reason, so this was an important concern at that time (spring 2006).

---

[3] DMO means the integration of real, virtual (man-in-the-loop), and constructive (computer generated) capabilities, systems, and environments. For example, linking high-fidelity simulators at different locations so that pilots can fly with each other in a simulated environment would be an example of DMO.

[4] Marken et al., 2007, provides additional information on advanced simulator training, as well as on this AFSO-21 initiative.

Our model runs showed that the proposal could have the desired effect but at a considerable cost to the units. Manning of the F-15C fleet, which had been projected to rise above 125 percent of authorizations by 2009 in the absence of simulator credit, would rise only slightly above 105 percent in its presence. Inexperienced pilots, whose flying was projected to decline from eight sorties per month in 2007 to fewer than six by 2012, would fly ten or more sorties per month. And these ten sorties per month would be flown in actual aircraft. Simulated sorties would be extra.

The cost of these improvements, however, is that they limit opportunities for pilots to serve second-tour operational assignments in their primary aircraft. In the cases that did not include the simulator credit, our model estimated that about 100 second-tour pilots would have to be assigned to operational F-15C units in FY 2009 and later years. With simulator credit in place, 25 to 30 percent fewer second-tour pilots would be assigned, until the production of first-tour pilots finally bottomed out in FY 2012. The change in the experience definition had resulted in pilots being counted as experienced earlier in their first operational tour, which meant that at any point in time there would be more experienced first-tour pilots. Since Air Force Personnel Center (AFPC) policies have the effect of holding constant the number of experienced pilots in operational units,[5] the increase in experienced first-tour pilots reduced the API-1 billets available for second-tour pilots.

---

[5]   The number of first-tour pilots in operational units has been determined by the production rates in recent years, and the number of these pilots who are experienced is determined by the number of hours they have been able to fly. Given these quantities, AFPC assigns enough second- and third-tour pilots to operational squadrons (a) to bring the total manning up to *at least* the authorized manning and (b) to bring the experience level up to at least a target experience level. The target experience level is set by policy, and is tantamount to setting a lower limit on the number of experienced pilots in operational units. For a decade or more, operational units have always had large numbers of inexperienced first-tour pilots, and by the time criterion (b) (the experience level) has been met, criterion (a) (the manning level) has been surpassed.

Discovery of a Potential Second-Tour Choke Point

A second operational tour (or its equivalent) in a pilot's primary weapon system is key to his later career development,[6] and it typically follows an intervening ALO–forward air control–AETC (ALFA) tour.[7] We therefore modified our model to track second-tour opportunities in the primary weapon system so that we could quantify these results for our Air Staff and ACC sponsors.

Runs with the new version of the model showed that there was indeed a looming shortage of second-tour billets. A typical F-15C pilot completing an ALFA tour in the second half of FY 2010 would have to wait almost four years before a slot would become available in an operational unit. This gap in their flying careers would inhibit their operational development.

We calculated that, in the cases with simulator credit, keeping pilots from waiting for their second tours would require more than 90 alternative second-tour billets at the end of FY 2010 in addition to the 77 second-tour billets in operational units. Significantly, the problem loomed even for the cases that used the old definition of experience, although the projected waiting times were shorter and the number of alternative billets needed to avoid waiting was lower. Lower production numbers did mean that fewer candidates completing ALFA tours were competing for larger numbers of second-tour assignments.

---

[6] The need for a second-tour opportunity in the primary weapon system was the essentially unanimous consensus requirement for later supervisory or staff expertise among both incumbent staff officers and operational supervisors that we interviewed during an earlier phase of our analysis (Marken et al., 2007). Essentially equivalent second-tour experience is available in nonoperational flying assignments, such as FTU instructors or aggressor pilots. Relatively few such assignments exist, however, and these usually follow an operational flying tour without an intervening ALFA assignment. We did not address them in our model runs, except for calculating the number required to avoid extensive waiting periods for second-tour assignments.

[7] ALO assignments with the Army are still an important ALFA component, as are AETC instructor assignments flying trainer aircraft, but the forward air control assignments were subsumed in the OA-10 following the retirement of the OV-10 aircraft and have been replaced in the ALFA world by UAS assignments. The ALFA acronym itself remains in widespread use despite these modifications.

We were receiving contemporaneous real-world confirmation of the validity of these results from AFPC. The chief of F-15C assignments reported that, for the first time during his tenure, he had been unable to fill all his F-15C recurrency course slots during the fall assignment cycle because the flying units lacked sufficient billets for the pilots' essential follow-on assignments after completing the course.[8]

Thus, the absorption choke point (not enough room in a unit to absorb enough new, inexperienced pilots) was effectively being moved to a second-tour choke point (not enough room for second-tour pilots) by the change that allowed credit for simulator time in the experience definition. During the post-Vietnam drawdown, the rated supplement, comprising only nonrated support billets, had been created to provide opportunities for rated officers to begin new career tracks in nonrated specialties, if they so desired. Those who were forced to wait for a flying assignment to continue their operational development to qualify to serve on the staff of a major command (or higher) or in an operational supervisory (or command) billet were often required to take an intervening assignment in what could be called an "API-0" billet.[9] Our model runs therefore indicated that all F-15C pilots completing an ALFA tour (which is the majority) after FY 2010 would be required to interrupt their operational development with at least a four-year API-0 period before they could return to flying.

Operational Units Require Second-Tour Pilots for IPs and Flight Leads

It was clear that reducing second-tour billets in operational units would have detrimental effects on pilots' operational development, but what effect would it have on the units themselves? To address this question, we counted all API-1 and API-6 pilots grades O-5 and below

---

[8] Note that the corresponding fall assignment cycle would mark the beginning of FY 2007.

[9] Aviation ratings were awarded only to pilots and navigators during the post-Vietnam period. They now include air battle managers as well. The actual term at that time was "rated position indicator 0" (RPI-0) because the indicators were then used only with rated billets. They were subsequently expanded to include all aircrew billets, including career enlisted aviators.

assigned or attached to operational F-15C squadrons. Assignment policies and practices at the time ensured that all pilots were exiting their initial operational assignment at very close to 500 hours of PAI flying time. Pilots with more than 600 PAI hours were therefore definitely on their second or subsequent operational tour, and pilots with more then 400 PAI hours were either in the final year of their initial operational assignment or on a second or subsequent operational tour. Minimum PAI hour requirements for upgrade were 500 hours to become an IP and 300 hours to become a four-ship flight lead. Studying the credentials and flying hours of pilots in the sample yielded some compelling results:

- Fewer than 5 percent of assigned and attached F-15C pilots with under 600 PAI hours were credentialed as IPs.
- Fewer than 13 percent of assigned and attached F-15C pilots with under 500 PAI hours were credentialed as four-ship flight leads.
- Fewer than 4 percent of assigned and attached F-15C pilots with under 400 PAI hours were credentialed as four-ship flight leads.

We concluded that pilots on second (and subsequent) tours were absolutely indispensable in the operational units to provide the IPs and flight leads required for essential in-flight supervision and the necessary upgrade training critical for the units to continue to sustain themselves over time. A limited number of pilots, typically FAIPs, might become four-ship flight leads during the final year of an initial operational tour, but for several years, virtually no one has been able to become an IP during an initial tour.[10]

These second-tour insights also showed that, if second operational tours are important for full development of operational competencies, the Air Force would not be able to fill unmanned staff positions competently simply by changing the definition of an experienced pilot.

---

[10] The data were collected and provided by ACC Air and Space Operations, Flight Operations Division (ACC/A3T). We briefed the Deputy Chief of Staff for Air, Space and Information Operations, Plans and Requirements on these results and the corresponding model runs on 26 June 2006 and gave the same briefing to ACC/A3T the following week. He, in turn, forwarded our slides and the briefing particulars to ACC's Director of Air and Space Operations.

Effects of Related Cuts in Flying Hours

As a result of DoD-directed budget reductions, the Air Force decided in FY 2006 to decrease its total flying hours by 10 percent.[11] Because the number of hours in formal training programs and ongoing operational test and evaluation programs could not be reduced, the flying hours available for operational fighter units had to be reduced by approximately 15 percent to achieve an overall cut of 10 percent.

In the F-15C, these flying hour cuts largely offset the gains from crediting simulator sorties toward experience. As mentioned above, crediting simulator sorties toward experience in F-15C squadrons can lower the manning from 125 percent of authorization to under 105 percent by the end of FY 2010 (by reducing the number of second-tour pilots needed to maintain a minimum experience level). With both the simulator credit and cuts in flying hours, the manning rises to 120 percent of authorizations. Similarly, crediting simulator sorties toward experience has the ultimate effect of increasing flying by inexperienced pilots from about six to as many as ten live sorties per month in the latter half of FY 2011. Cutting flying hours drops it to under five live sorties per month.

The flying-hour cuts that AFSO-21 directed affected CAF units much more than MAF units because AMC aircrew managers were again able to offset most of the cuts in O&M hours with corresponding increases in TWCF hours. These cuts therefore especially affected fighter and bomber units, and the former represented a sizable majority of the total CAF hours. The cuts were taken in every year of the FYDP associated with the FY 2008 program objective memorandum (POM). As this monograph was in preparation, additional cuts were expected in conjunction with the FY 2010 POM process. This was emerging despite concerted CAF efforts to avert additional cuts in flying hours. CAF was concerned because projected efficiencies attributed to AFSO-21 initiatives in the intervening execution years could not be realized.

---

[11] According to Wicke, 2006, the reductions would apply each year from FY 2008 to FY 2013.

The TAMI 21 Task Force: Proposals to Improve Aircrew Management

In August 2006, as F-15C and F-16 units continued to struggle with manning and experience problems, the Air Force Deputy Chief of Staff for Air, Space, and Information Operations, Plans and Requirements chartered the TAMI 21 Task Force and invited us to participate and provide analytic support. The purpose of the task force was to

- develop solutions to major aircrew management issues involving experience, training, and absorption
- collect and assess background information, as required
- present recommendations to senior leadership for action.

The task force was put on a fast track, with its first meeting set for October 2006 and its objectives to be met by March 2007. The participants were told to examine aircrew management using "a 10 year sight picture," and one goal of the task force structure was to "ensure objective discourse, analysis, and proposal development to underpin effective aircrew management."[12]

Challenges related to aircrew management had continued to build throughout FY 2006. Despite the increased flexibility that the redefinition of *experience* gave for moving first-tour pilots out of the operational F-15C units, the F-22 transition under way at Langley AFB exacerbated the manning and experience issues. The transition effectively shut this location down to incoming F-15C B-course graduates and therefore increased the flow of new pilots to the other F-15C locations. These increases in the flow of new pilots especially hurt single-squadron locations (no locations have more than two squadrons), with manning levels at Mountain Home AFB, for example, exceeding 120 percent by September 2006 with little prospect for improvement. Elmendorf AFB

---

[12] The quotes and stated purpose are taken from the original package establishing the task force, which was transmitted by Air Force Operations, Plans, and Requirements to the major command vice commanders; Air Force Manpower, Personnel, and Services; the AFPC commander; and the PAF Director on August 30, 2006. The package also confirmed that the task force chairman would be the Director of Current Operations and Training and its working chairman would be the AF/A30-AT chief.

would soon begin its own transition to the F-22, thereby continuing to aggravate overmanning issues in other locations. Adding to the problem, the F-22 was not an absorbable system and accepted only experienced fighter pilots. There have been sequential delays in implementing an F-22 B-course, and the first pilots are not currently projected to graduate until sometime in FY 2009.[13]

Overmanning and related issues were also looming in F-16 units because the Base Realignment and Closure (BRAC)–directed closure of five operational squadrons was to be implemented and completed in FY 2007. This was earlier than originally programmed and meant that these units—three squadrons at Cannon AFB and one each at Mountain Home and Eielson AFBs—were also closed to B-course graduates, thus increasing the numbers being sent to the other F-16 locations. As a result, the manning level at Misawa Air Base, for example, would increase from a manageable 105 percent in September 2006 to 126 percent in January 2007, reaching a distinctly unmanageable 141 percent by May 2007.[14]

We were able to predict the manning increases that were occurring in both the F-15C and the F-16 with our dynamic model runs during the course of the TAMI 21 sessions. We will next examine some of those model runs.

Model Results Used in Initial TAMI 21 Discussions

The results described below are recorded through the end of FY 2016. These model runs include the AFSO-21 policy decisions to incorporate simulator credit and cut the available flying hours throughout the FYDP. The task force points of contact supplied force-structure information through FY 2011 and for FY 2016, but we had to interpolate to estimate fighter force structure values (PMAI) for the intervening years (FYs 2012–2015). In addition, the existing new-pilot production

---

[13] Manning figures are from AFPC Deputy Personnel Assignment Operations (AFPC/DPAO). Other information sources include AF/A3O-AT and ACC Air and Space Operations' Flight Operations Division (ACC/A3T).

[14] Manning figures are from AFPC/DPAO.

and distribution data were available through FY 2013, but we had to extrapolate from these to obtain numbers for FYs 2014–2016.[15]

Although we obtained results for every absorbable operational fighter MDS, we focused on those for the F-15C and F-16 because these systems had been identified earlier as being on the verge of breaking. The default runs were intended to replicate, as closely as possible, the conditions that would occur in the operational units with no modifications to the Air Force policies in effect at that time. We will present these results first.

Projections of the Consequences of Doing Nothing

Figure 4.1 illustrates how maintaining the aircrew management policies that were in effect in December 2006 would affect unit manning for F-15C and F-16 units.

F-16 units started at about 105 percent manning at the end of FY 2005, and manning levels rise fairly steadily to 140 percent by 2013. This is a consequence of continuing to send high numbers of new pilots to F-16 units while making drastic cuts in the number of F-16 aircraft. The F-16 PMAI falls from 414 aircraft in 2006 to 342 in 2007 (a 17-percent drop), where it remains until 2012, while programmed entries into F-16 training do not decrease proportionally.[16] Indeed total fighter production numbers had to be reduced in FYs 2008–2009 to accommodate a BRAC-directed change in the location of the formal IFF lead-in training course, required of all new fighter pilots. The total production numbers for all fighters increase rapidly thereafter to meet the Air Force objective of returning to an annual production rate of 330 fighter pilots by the end of the then-current FYDP in FY 2013.

The manning trend is similar in the F-15C, but the production numbers are more cyclic, with the early decrease exhibiting the con-

---

[15] We used ACC/A3T estimates of when the advanced, DMO-capable simulators would become available in the various fighter units. Force structure information was provided by ACC/A5B and AF/A3O-AT. Pilot production and distribution information came from AF/A3O-AT and was vetted through AETC/A3R to ensure that the pending reductions in numbers of fighter FAIPs were accurately incorporated in the runs.

[16] From FY 2012 to FY 2016, the PMAI drops from 342 to 168. Recall that API-1 manning authorizations are directly determined by PMAI.

Figure 4.1
Effects of Rated Management Policies Existing in December 2006

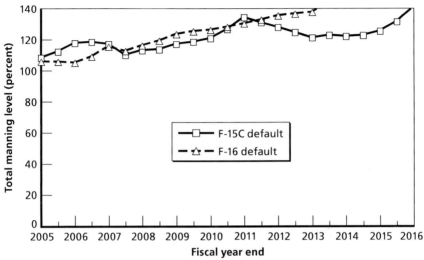

RAND *MG855-4.1*

certed efforts of the assignment system to bring manning levels back under control following the change in the definition of experience in 2006 and the later decrease reflecting the uncertainties in the timing and quantities of F-15C unit conversions to F-22s after FY 2011. In any case, F-15C manning levels exceed 120 percent in FY 2010 and remain at least that high thereafter. For both aircraft, fleetwide unit manning after 2010 exceeds the overmanning that existed at Pope AFB in July 2000 and which, as we noted earlier, led to degradation of pilot skills.

With more people in the units than are authorized for the number of aircraft, it is difficult to distribute flying hours equitably among experienced and inexperienced pilots. Most sorties inexperienced pilots fly must be under the supervision of an experienced pilot, with either flight lead or IP credentials, flying in the same flight in another aircraft (although occasionally the supervisor is an IP accompanying the inexperienced pilot in a two-place fighter). Thus, if there are too many inexperienced pilots in a unit, each tends to fly less often, while experienced flight leads and IPs may fly more than required to maintain their own personal proficiency.

Figure 4.2 shows the projected training available to the inexperienced pilots in terms of SCM. SCM for inexperienced F-16 pilots declines steadily after 2007, going below four in 2013; the desired SCM rate for inexperienced pilots is 12, with a minimum of ten to maintain CMR status under the RAP training requirements and to meet the experience criterion during the initial operational flying period. This minimum rate could presumably drop to eight when up to 20 percent simulator credit is allowed under RAP and in the revised experience definition. In keeping with the manning cycles, the F-15C SCM rates show more variation but still drop below five in 2011 and stay at six or below thereafter.

These SCM rates are below the average of about 5.5 that existed at Pope AFB in July 2000. This value means that new F-15C/F-16 pilots will leave their first fighter tours after 36 months with only 280 to 300 flying hours.[17] This is not enough to become experienced, as Figure 4.3 makes clear.

The vertical axis of Figure 4.3 is the time required in months for an SUPT graduate to become experienced. After 2010, the lines for both F-15C and F-16 pilots become dashed lines leveled out at 36 months (the standard tour length and the time at which a pilot will be transferred to a new assignment). What this means in terms of the model is that neither F-15C nor F-16 pilots can become experienced in a 36-month tour after 2010—they will be transferred to their follow-on assignments before they have flown enough hours. Note that this is under the relaxed definition of experienced, which counts 100 hours of simulator time toward the 500 hour total.[18] Figure 4.4 shows the corresponding reduction in unit experience levels.

The vertical axis in this figure shows the percentage of assigned API-1 pilots in the units who are experienced. The proportion of expe-

---

[17] New pilots typically fly nearly 80 hours in their primary mission aircraft during B-course training. The programmed ASD in F-15Cs is 1.3 hours, and in F-16s is 1.4. With 4.3 sorties per month and after a 36-month tour, an F-15 pilot would have 80 + (36 × 4.3 × 1.3) = 281 hours. Using the same SCM for F-16s gives 297 hours.

[18] Note that the models gave similar credit starting in FY 2008 for the F-15C units and the Block 50 F-16 units. The Block 40 F-16 units, however, were not programmed to get advanced simulators until FY 2011, so the credit was phased in accordingly.

Figure 4.2
Training Sorties for Inexperienced Pilots Under Default Policies

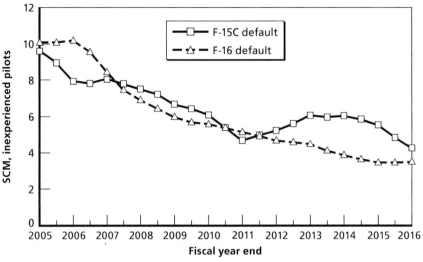

Figure 4.3
F-15C and F-16 Pilots Becoming Experienced Under Default Conditions

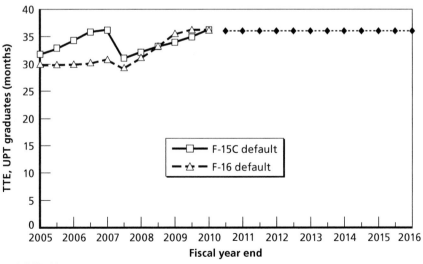

Figure 4.4
Experience Levels in F-15C and F-16 Units

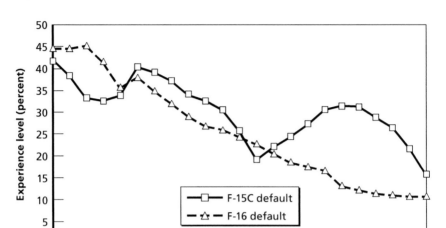

rienced pilots in F-16 units drops below 35 percent during FY 2008 and continues to drop thereafter. F-15C experience levels stay below 35 percent after 2010.

There were critical in-flight supervision issues at Pope AFB in July 2000 even when the aggregate experience level was over 36 percent, so the experience levels the model predicts have troubling implications for the health of F-15C and F-16 units.

As a final example of the state of fighter aircraft units under the aircrew management policies in effect in December 2006, Figure 4.5 shows the percentage of API-1 positions filled by pilots who are in a first tour. This is an important measure because second-tour pilots returning to operational flying following an ALFA tour must normally also fill API-1 billets until they have flown enough to acquire the qualifications required to upgrade to flight lead and/or IP status.

This measure reaches 80 percent for F-16 units in FY 2007 and remains at least that high thereafter, going above 100 percent in

Figure 4.5
First-Tour Pilots Fill More Than 80 Percent of the API-1 Authorizations

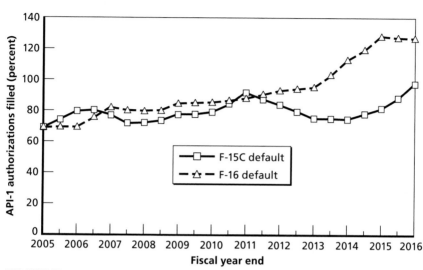

RAND *MG855-4.5*

FY 2014.[19] Except for the manning cycles that we have already discussed, it also remains above 80 percent in F-15C units, and we think that the Air Force is likely to adjust distribution rates as required to bring these fill rates into a better balance.

The results presented in Figures 4.1 through 4.5 have unacceptable consequences for training and aircrew development. We have addressed only the F-15C and the F-16 because these systems faced the most imminent problems at the time. The other absorbable fighter systems actually faced similar issues that would occur at later points. Table 4.2 summarizes the results for all the fighter systems that should be absorbable within the TAMI 21 analysis time frame.

The default model runs for the TAMI 21 Task Force indicate that every absorbable fighter system breaks by FY 2015. This is primarily the result of trying to force too many new fighter pilots through the training system while aircraft inventories continue to decline and further

---

[19] This measure can exceed 100 percent in the model because of overmanning: There are more pilots in the unit than are authorized.

Table 4.2
The Default Model Runs for the TAMI 21 Task Force Indicate That Every Absorbable Fighter System Breaks by FY 2015

	FY 2005	FY 2006	FY 2007	FY 2008	FY 2009	FY 2010	FY 2011	FY 2012	FY 2013	FY 2014	FY 2015	FY 2016
A/OA-10	Hlth	Hlth	Hlth	Hlth	Hlth Str	Str	Str Brk	Str	Str	Str	Str Brk	Str Brk
F-15C	Str	Str	Str Brk	Str	Str	Str Brk	Brk	Brk	Brk	Brk	Brk	Brk
F-15E	Hlth	Hlth	Hlth Str	Str	Str	Hlth	Str	Str	Str	Str	Str Brk	Brk
F-16	Str	Str	Str	Str	Str Brk	Brk	Brk	Brk	Brk	Brk	Brk	Brk
F-22				Hlth	Hlth	Hlth	Hlth	Hlth	Str	Str Brk	Brk	Brk

Healthy	Hlth
Stressed	Str
Broken	Brk

NOTE: More years and more MDSs are included here than were in Table 4.1 to accommodate the TAMI 21 charter to examine the issues with a 10-year outlook. Simulator credit is given starting with FY 2008 in the F-15C, Block 50 F-16, and F-22 (which, at the time, was projected to begin absorbing new pilots during that year). Credit begins in FY 2010, 2011, and 2012 for the F-15E, Block 40 F-16, and A-10, respectively.

limit the absorption capacities in the systems. Correcting this situation was the most fundamental recommendation the TAMI 21 Task Force made, despite recognizing that doing so would, over time, decrease the availability of fighter pilots to fill nonflying billets.

Options for Fixing Fighter Unit Problems Studied by TAMI 21

Working with future force structure projections from the Air Force, we used our model to estimate that maximum absorption of new fighter pilots would decline to about 200 by FY 2016. This value was remarkably close to the value AF/A3O-AT had determined independently using static models. Our model runs also portrayed the health of the active fighter units during the intervening years. The series of figures that follows exhibits unit characteristics when fighter pilot entries are reduced to yield healthier units. These results are, again, for the F-15C and F-16 units.

Figure 4.6 gives the manning levels. The only differences between the runs labeled "TAMI 21" and those labeled "Default" are the production and distribution numbers for new pilots. The default runs

Figure 4.6
Effects of Reduced Pilot Production on Manning Levels

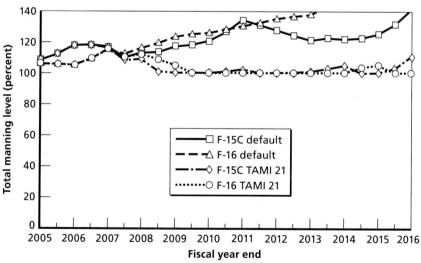

are the same as those depicted in Figures 4.1 through 4.5 and Table 4.2. The first quarter of FY 2007 was almost over when the runs were made, so only negligible adjustments could be made for that year, with somewhat limited reductions in FY 2008 as well, because many in that cohort were already in the undergraduate training pipeline. The adjusted total fighter production figure was 229 pilots in FY 2011 and reached 200 pilots in FY 2014. The manning level for both systems would be below 110 percent by the end of FY 2009 and remains under control thereafter.

Similar improvements occur in the training available for inexperienced pilots. Figure 4.7 shows this training, as measured in SCM, after reducing the flows of new pilots into the active units. The available training exceeds eight SCM in FY 2009 for both systems, which, when advanced simulator credit is included, will meet RAP training minimums and ensure that the pilots become experienced within the initial operational flying period.

Experience issues also improve dramatically once the excessive flow of pilots is corrected. Figure 4.8 indicates that pilots will meet the 32-month time-to-experience (TTE) objective in both systems in FY 2010 and beyond. The shorter TTEs help generate higher experience levels in the units. Figure 4.9 illustrates these experience levels. They very nearly meet or exceed the CAF objective of 55 percent after FY 2010. Indeed, F-16 experience levels exceed 60 percent from FY 2010 until FY 2016, indicating that this system would probably be able to absorb more new pilots in these years than we have projected. This represents an opportunity for an informed director of Air Force Operations to adjust annual pilot-production and distribution numbers appropriately to optimize the overall absorption capacity while ensuring continuing unit health.

The percentage of API-1 billets that first-tour pilots fill also decreases to acceptable levels when the flow of new pilots into the active units decreases. Figure 4.10 shows these fill rates, which remain near 60 percent (or below) after FY 2010. This is an acceptable first-tour fill rate for the API-1 authorizations. The TTEs shown in Figure 4.8 mean that more than 10 percent of the first-tour pilots, on average, can be experienced at any given time and that the overall experience

Figure 4.7
Improvements in Available Training for Inexperienced Pilots

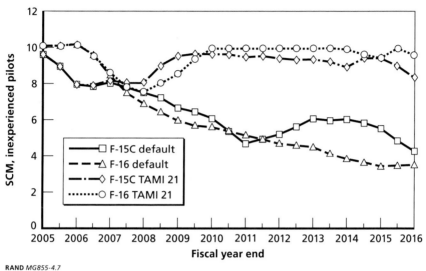

RAND *MG855-4.7*

Figure 4.8
The 32-Month Time-to-Experience Objective

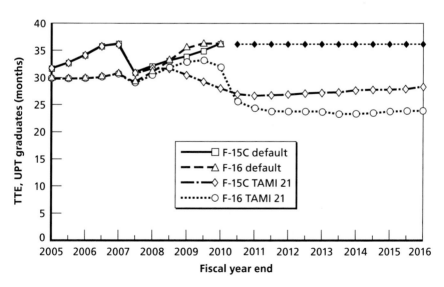

RAND *MG855-4.8*

Figure 4.9
CAF's 55-Percent Experience-Level Objective

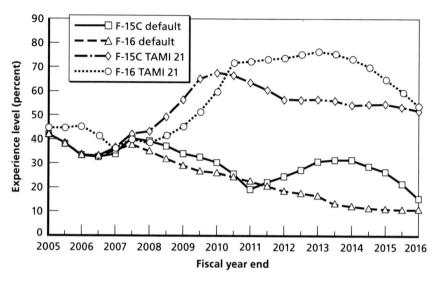

RAND *MG855-4.9*

Figure 4.10
First-Tour Fill Rates

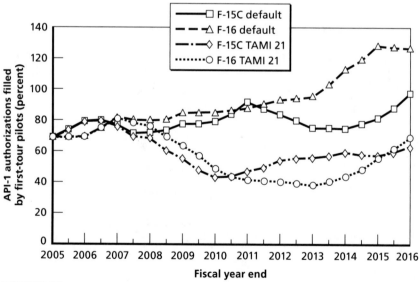

RAND *MG855-4.10*

level (obtained by adding this percentage to the 40 percent of second- and subsequent-tour pilots assigned to the API-1 billets) will exceed the 50-percent experience objective we used for this sequence of model runs.[20] These results illustrate what can be done to bring manning and experience issues under control. As Figures 4.6 through 4.10 show, these runs provide a much more acceptable training environment for the operational units.

The results for the fighter MDSs are summarized in Table 4.3. As the table indicates, the worst of the stressed conditions are over after FY 2009. The isolated stress period for the A-10 in FY 2011 could easily be avoided by adjusting the new-pilot distribution schedule to better match available absorption capacity. Also, the isolated yellow areas in FYs 2014 through 2016 should not present a serious problem because we feel they could be readily eliminated by distributing new pilots more judiciously across MDS to match the available absorption capacity, especially if the F-35 becomes absorbable during that period. These results were obtained simply by controlling the flow of new pilots into the active fighter units to avoid exceeding the absorption capacities of the units.

Recommendations of the TAMI 21 Task Force

The TAMI 21 Task Force recognized that lowering the production rate for new fighter pilots to 200 per year would not address the principal problem of filling existing and projected vacant staff billets. The group therefore examined a number of options for developing alternative sources for filling empty billets and/or options that would reduce existing requirements for nonflying aircrew members in staff positions and other related billets. The last category of options would prove espe-

---

[20] If we take a nominal value of 25 months for the average TTE depicted in the TAMI 21 runs in Figure 4.8 and use a 32-month average tour length, we see that 7/32, or nearly 22 percent, of the first-tour pilots will actually be experienced, on average. This equates to over 13 percent of the total API-1 authorizations, yielding an experience level in excess of 53 percent on adding it to the 40 percent of the API-1 pilots serving in their second (or subsequent) tour. The actual CAF experience objective has remained at 55 percent since the advent of the RDTM system in the 1970s, and we were asked to use 55 percent as the experience objective in subsequent model runs, which we willingly agreed to do. We would never argue that the 2-percent difference would not be within the expected margin of error for our models.

Table 4.3
Summary of TAMI 21 Directed Model Runs

	FY 2005	FY 2006	FY 2007	FY 2008	FY 2009	FY 2010	FY 2011	FY 2012	FY 2013	FY 2014	FY 2015	FY 2016
A/OA-10	Hlth	Hlth	Hlth	Hlth	Hlth	Hlth	Hlth Str	Hlth	Hlth	Hlth	Hlth Str	Hlth
F-15C	Str	Str	Str Brk	Str	Hlth	Hlth	Hlth	Hlth	Hlth	Hlth Str	Hlth	Hlth Str
F-15E	Hlth	Hlth	Hlth Str	Str	Str	Hlth	Hlth	Hlth	Hlth	Hlth	Str	Str
F-16	Str	Str	Str	Str	Str	Hlth	Hlth	Hlth	Hlth	Hlth	Hlth Str	Hlth
F-22					Hlth	Hlth	Hlth	Hlth	Hlth	Hlth	Hlth	Hlth

Hlth	Healthy
Str	Stressed
Brk	Broken

cially elusive because of increasing needs to fill aircrew slots at a growing number of AOCs, whose importance to the modern expeditionary Air Force was increasing, and to supply UAS operators for the rapidly expanding (and also increasingly important) UAS fleet.[21] However, an essential top-to-bottom review of all aircrew requirements was not feasible within the time available, so the TAMI 21 Task Force deferred it.

The task force concluded that achieving healthy units by curtailing the flow of new pilots into active fighter units to meet absorption constraints would be an essential component of any reasonable long-term solution. It also recommended the following alternatives:

- establishing an independent, self-sustaining UAS operator career field
- using the air reserve component (ARC) force structure to absorb and develop new active pilots effectively
- exploring the full potential of other manning alternatives for rated staff positions, such as
  - appropriately developed ARC personnel
  - career enlisted aviators (CEAs)
  - prior-military civilians with the appropriate military training, experience, and background
- allowing adequate flexibility for aircrew managers, under the Director of Air Force Operations, to make minor adjustments in the annual production numbers of new aircrews, as needed, to maintain the health of operational units.[22]

---

[21] April 2008 "redline" numbers provided by AF/A3O-AT indicated that UAS requirements could increase from 458 in FY 2008 to 1,060 by FY 2013. Note that the Air Force thinks of the fleet in terms of its operators and their equipment, rather than strictly in terms of the UAVs alone, which is why the more-encompassing *UAS* is now standard.

[22] These recommendations should not be confused with the TAMI 21 initiatives that would eventually be publicized by the Air Force in May 2007. We will discuss the evolution of the differences in a subsequent section. All the information in this subsection is taken from the personal notes of one of the authors, who attended every TAMI 21 Task Force meeting session.

The first two recommendations can be roughly quantified. The best estimate available at the time for the annual requirement of new UPT graduates entering a dedicated UAS career field was 50 to 100 pilots per year. An Air Force study conducted by AETC and the UAS Operations Group at Creech AFB concluded that these UAS operators would develop career skills very similar to those of an A-10 pilot, with certain combat experience deficiencies easily offset by advantages gained from a better understanding of the intelligence analyses and communications needs required within the tactical air control system when conducting close air support operations (Clary, 2006, slide 21). Thus, these officers would be fully qualified to fill many of the nonflying staff billets that required fighter pilots and consequently remained unfilled because inventories were inadequate. Moreover, the Air Staff's FY 2016 force structure information that TAMI 21 received indicated that about 36 percent of the 991 potentially absorbable fighter aircraft in the total force would belong to the guard and reserve in 2016. If these 361 airframes could absorb 50 to 75 new active pilots per year, the total number of fighter pilots and UAS operators that could be absorbed would increase to 300 to 375 pilots per year, providing a more robust inventory that potentially could meet the original objectives of the Air Force leadership. We recommend this as a total force integration (TFI) objective (see discussion in Chapter Five).[23]

In December 2006, the chairman of the TAMI 21 working group briefed the Chief of Staff of the Air Force (CSAF) on the recommendations above. CSAF, however, accepted few of these. In guidance received after the presentation, aircrew managers were directed to accelerate the return to the 1,100 and 330 UPT production goals. Shortly after the presentation, ongoing experiments to train nonpilots as UAS operators were directed to terminate early. Encouragement to

---

[23] The information on career expectations for UAS operators is from the study report briefing (Clary, 2006). The annual figure of 300 to 375 absorbable pilots is the sum of the 200 fighter pilots that can be absorbed in the active units, the 50 to 100 new UAS operators that would be required, and the 50 to 75 additional active pilots that would be absorbable in the remaining guard and reserve units. Our model runs also indicated at that time that the upper bound could increase to about 395 per year if the AFSO-21 directed flying hour reduction were to be reversed.

pursue the TFI initiatives and the other alternative-manning options continued, but the TFI initiatives, once regarded as the most promising of these, still experience significant implementation problems that will be addressed in Chapter Five. The remaining alternative-manning options, although important, were and continue to be inadequate for resolving the existing issues.[24]

Other Events and Decisions That Followed TAMI 21

The Rated Sustainment Working Group

Following the reaffirmation of the CSAF-directed 1,100 and 330 pilot UPT production goals, the Air Force Deputy Chief of Staff for Manpower and Personnel directed his Rated Force Policy Branch Chief to chair a working group to examine ways that the Air Force might manage an annual pilot production of 1,100 total pilots and 330 fighter pilots without damaging mission accomplishment and officer career development. The dual focus of the working group drew the full support of the Deputy Chief of Staff for Air Force Operations, Plans and Requirements, who directed the full participation of the Aircrew Management Branch of AF/A3O-AT. As a result, the working group membership included many of the working-level members of the TAMI 21 Task Force. RAND was asked to participate and continue to provide analytic support.

The tasking responded initially to a CSAF comment about reexamining the efficacy of a *rated supplement*, a term that the Air Force first used to describe the disposition of excessive numbers of rated officers who remained on active duty following the post-Vietnam drawdown, many of whom were reassigned into nonrated career fields. That term had, however, taken on a somewhat pejorative sense among members

---

[24] Most of the details in this and the previous paragraph were taken from the task force chairman's summary notes of his meeting with CSAF on December 18, 2006. The notes were distributed to all TAMI 21 participants. Some of the information was extracted from later clarifications of CSAF intent, as well as discussions before and during the task force's last meeting, in January 2007, in preparation for which the task force members received copies of the slides the chairman used during the presentation.

of Congress and analysts at GAO during the post–Cold War drawdown, so it was rejected in favor of *rated sustainment*.[25]

Not surprisingly, the conclusions of the Rated Sustainment Working Group were consistent with the findings of the TAMI 21 Task Force and fully in line with the supporting analyses that were part of the TAMI 21 effort. The basic conclusion was that, if the 1,100 and 330 objectives were pursued, the force structure would not be able to absorb all the new pilots. The best thing to do with the excess pilot-training graduates would be to establish a permanent bank of SUPT graduates who would not be expected to fly at any time in the future. The working group was in general agreement that it would be far better not to have trained these pilots in the first place because of the issues that would develop in nonrated accessions as the Air Force adhered to an existing officer end-strength constraint. Another preferred alternative would be using the excess pilots as UAS operators. Unfortunately, both options had been precluded from consideration during the interpretation and discussion period following the December 2006 TAMI 21 presentation to CSAF.

The Four-Star Conference

In March 2007, we briefed the Deputy Chief of Staff for Air Force Operations, Plans and Requirements as part of his preparation for the CSAF Four-Star Conference, scheduled for later that month. Our presentation focused on describing the consequences of a continued Air Force effort to produce more new pilots than the capacities to absorb and develop them would allow, given projected force-structure limits for the various weapon systems. We addressed the issues associated with the operational CAF, SOF, and MAF units, in each case examining problems with manning, training, and aging young aircrews, as well as any issues developing within the units. Our basic conclusions were as follows:

- Excess pilot pools must be removed from many operational MDSs to resolve serious operational issues. Current distribution-plan

---

[25] See, for example, GAO, 1997.

production goals will exceed active absorption capacity by 150 to 200 pilots per year throughout the FYDP.

- *No* active operational MDSs have the capacity to accept diverted pilots to ease mission stresses in affected units. Several systems may soon require return-to-fly boards for pilots exiting ALFA or staff tours.
- The Air Force has two options for controlling excess pilot numbers:
  - reducing the input of new pilots entering active units
  - moving excess pilots out after entry.

We urged the Air Force to reconsider its decision against developing dedicated UAS operators because the plan to assign only pilots who had been previously mission-qualified in manned systems to such operations was not sustainable then and could only get worse as the ratio of unmanned-to-manned systems increased. We also suggested once again that the Air Force reexamine its TFI initiatives to address the dual objectives of (1) improving mission accomplishment in an expeditionary mode and (2) improving the absorption and development of new aircrews. We demonstrated how these initiatives could contribute to achievement of fundamental Air Force objectives.

Prior to the conference, we made additional model runs to examine the potential contribution that removing *inexperienced* pilots from their operational fighter units early would have on the manning and experience levels within these units. These pilots are also called "limited-experience" pilots. These analyses confirmed that, while moving limited-experience pilots out of operational units before the end of the normal initial operational flying period could improve the training environment in these units, the conditions required to achieve any measurable results seemed to be fairly dire. It would mean moving at least 25 percent of every entering UPT cohort in F-15Cs and F-16s throughout the FYDP, for example, after only 18 to 24 months of operational experience. These pilots would exit with between 280 and 300 flying hours in their primary fighters. Similar proportions of co-pilots would also need to be moved to correct overmanning conditions in bombers (and possibly SOF and CAF AC/MC/EC-130 systems). All these pilots could be reassigned to meet growing UAS demands and increasing SOF mission requirements for nonstandard aircraft systems, but none would be able to return to their original weapon systems until the flows of new pilots into them had been decreased enough to ease the absorption capacity constraints, which are primarily the result of the dwindling active force structure. We did not regard the permanent movement of limited-experience pilots from fighters and bombers into other weapon systems to be a viable long-term solution to unit overmanning problems because the hours these pilots fly are then lost and cannot be used to absorb and develop the necessary inventory of more-senior, staff-qualified CAF officers, who have been in such short supply.[26]

The conference restored two of the original initiatives of the TAMI 21 Task Force when it agreed to reduce the flows of new pilots into weapon systems where production goals exceeded absorption capacities and gave the director of current operations and training limited authority to manage pilot-production totals within the range of 950 to 1,050 new pilots per year.

Decisions That Followed the 2007 Four-Star Conference

Air Force leadership sought to simultaneously address right-sizing the overmanned fighter and bomber units and the growing need for more UAS operators and SOF pilots by moving limited-experience fighter pilots and bomber copilots directly into the new UAS and SOF billets while reducing the flow of new pilots into fighter and bomber units. This decision was announced in May 2007, along with other Air Force policy initiatives intended to address several aircrew management issues (Randolph, 2007). Other initiatives included

- eliminating CAF overmanning in operational units
- opening previously restricted airframes to new SUPT graduates
- using CEAs in nonflying rated requirements

---

[26] Moving limited-experience pilots from other fighters would not be required until FY 2011. Our model runs indicated that they would need to occur as follows: A-10 in FY 2011 and the F-15E and F-22 in FY 2013.

- using the total force to help absorb new pilots and provide staff experience
- increasing the minimum number of SCMs for inexperienced pilots
- ensuring that aircrew training requirements meet combatant commander needs.

Except for increasing the minimum SCM for inexperienced pilots, these initiatives were consistent with the working group's recommendations, and the Air Force publicized them as the TAMI 21 initiatives. The conference decisions that addressed eliminating overmanning in operational units were deemed the most critical, so two decisions were quickly implemented. First, the annual pilot-production goal would be reduced, from 1,100 to between 950 and 1,050, allowing some flexibility for aircrew managers to respond to increasing (or decreasing) needs for pilots. Second, some fighter pilots with limited experience (more than one year on station, but fewer than 400 fighter hours) would be transferred from fighter units to Air Force Special Operations Command (AFSOC) units (where they would remain as AFSOC assets for the rest of their careers) or to UAS positions (from which a limited number would someday return to fighter units). Volunteers for these moves would be sought first, but nonvolunteers would be reassigned if necessary. A total of about 80 pilots were to be transferred over the four assignment cycles from October 2007 to January 2008. In addition, 60 experienced F-16 pilots were to be moved to rated billets that would not involve flying F-16s (such as staff jobs and AETC IP positions) after completing their tours.[27]

As of December 2008, the status of the remaining initiatives was as follows:

- Only a few handpicked SUPT graduates had entered the F-22 B-course.

---

[27] About 40 inexperienced bomber pilots were also to be transferred to AFSOC and UAS billets.

- While AMC was able to convert several dozen rated officer billets to CEAs, the CAF major commands (ACC, Pacific Air Force, and U.S. Air Forces Europe) converted a total of two billets among them.
- The use of the total force in addressing absorption and staff manning issues has been ineffective (see Chapter Five).
- We understand that the proposal to increase RAP requirements for inexperienced pilots was rejected because of the associated cost estimates.
- Training activities, especially exercises, have stressed the scenarios that are encountered in operations Enduring Freedom and Iraqi Freedom.[28]

Predicted Consequences of the Decisions

We analyzed the potential consequences of the policy decisions that came out of the CSAF Four-Star Conference and the TAMI 21 initiatives. As a result, our subsequent model runs incorporated a slightly revised active future force structure; a significantly reduced pilot-production and distribution plan, in which future annual fighter-pilot production did not exceed 265 pilots per year; an adjusted experience objective of 55 percent; and a new maximum manning goal of 105 percent.

Removing excess pilots over four assignment cycles, a period of 16 months, coupled with the lower distribution plan did delay the projected onset of serious problems until FY 2013 for both F-15Cs and F-16s, but a long-term solution would require reducing the number of new pilots headed into the active units even further. To maintain unit health, the number of new pilots entering the active fighter units still had to decline to 200 per year by FY 2016. Figure 4.11 compares the Air Force distribution plan with our proposed production and distribution numbers over the period of interest.

The number of FAIPs is the same for both distribution plans, and the difference between the shaded and unshaded bars represents the reductions in IFF entries required to bring the operational fighter

---

[28] According to AFPC and AF/A3O-AT.

Figure 4.11
Comparing the Numbers of Pilots Entering Introduction to Fighter Fundamentals by Fiscal Year Under Two Plans

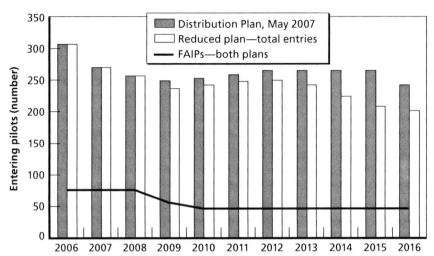

SOURCE: AF/A3O-AT provided the Air Staff pilot production and distribution plan, Distribution Plan Version 4.1, May 2007. The reduced plan was developed using RAND model runs.
RAND *MG855-4.11*

squadrons to manning levels between 100 to 105 percent, raise experience levels above 55 percent, and increase the number of sorties inexperienced pilots fly per month to about ten. These model runs were made in June and July 2007, with FY 2007 almost three-quarters complete and with insufficient lead time to influence pilot-production rates for FY 2008. The IFF entries are therefore identical through FY 2008. We tracked IFF entries, rather than B-course graduates, because production goals are always based on UPT graduates and do not include attrition that could occur before pilots report to their initial operation assignments.

The Removal of *Experienced* Pilots Has Little Effect
As noted above, the Air Force's May 2007 policy decision included moving 60 experienced F-16 pilots completing their tours from flying units to other rated assignments. Some Air Force managers thought that this component of the policy would make extra fighter pilots

available for staff positions that required rated personnel. However, the removal of these experienced F-16 pilots from flying units provides no significant benefit beyond what early transfer of inexperienced F-16 pilots offers.

As explained earlier, aircrew management policy assigns enough second- and subsequent-tour pilots to the units to maintain constant numbers of experienced pilots. During the four assignment cycles over which the moves occurred in our model, it shows that there are no extra experienced pilots. There are just enough to make the "book" experience level equal to 55 percent. So, if an experienced pilot leaves the unit, the model immediately replaces that pilot with another.[29]

UAS Career Field Revisited

In February 2008, CSAF appeared to retract his earlier opposition to sending UPT graduates directly to UAS systems and to an independent UAS career field. As reported in the press,

> Gen. Michael Moseley, Air Force Chief of Staff, . . . has directed a number of bureaucratic and organizational changes to raise the profile of unmanned aerial vehicles within the service. First, . . . he instructed the USAF Weapons School at Nellis AFB, Nev., to stand up a UAV squadron by July. He also instructed Gen. William Looney, head of Air Education and Training Command, to begin looking into assigning airmen directly to UAV operations as first assignments. Moseley said the details of such assignments are still to be worked out, but one possibility would be for fighter- or bomber-track pilots to serve a two-year assignment first with UAVs. (To offset a current shortage of UAV pilots, the service reportedly already has tapped more than a hundred experienced fighter and bomber pilots for UAV duty.) Finally, Moseley said the service will assign a new Air Force Specialty Code to UAV operators and sensor operators. These airmen would then have both a "primary" and a "secondary" AFSC to better capture

---

[29] Recall that the "book" experience level is the experience level reported by the Air Force, which uses as the denominator the number of *authorized* pilots in a unit. When overmanning occurs because of excess inexperienced pilots, the actual experience level is lower than the "book" level.

their UAV bona fides. These moves are a reflection of the growing importance of UAVs within the Air Force's operations. Moseley noted that the Fiscal 2009 budget funds 93 new Air Force aircraft—and 52 of them are unmanned Predators, Reapers, and Global Hawks.[30]

He remained ambiguous, however, about whether these pilots might expect to fly another manned system later. Indeed, the article cites the possibility of fighter- or bomber-track pilots first serving a two-year UAV assignment. But it does not confirm this requirement, which would be extremely difficult to execute because of the perishable nature of the flying skills of new UPT graduates, requiring extensive refresher training to fly fighters or bombers after a two-year hiatus. Indeed, the pilot banking experiment the Air Force conducted after the post–Cold War drawdown eventually required returning pilots to refly about 60 percent of the T-38 training they had received in UPT before progressing to fighters or bombers.[31]

This issue was discussed at a rated management conference, cochaired by the director of Air Force Operations, Plans and Requirements, and Air Force Manpower, Personnel, and Services on February 28, 2008, but it was not completely resolved until after a Plans and Requirements conference in May 2008. At that time, CSAF approved the conference minutes, including implementation of the policy of sending UPT graduates directly to UAS billets with no guarantee of ever returning to manned systems.[32] While this decision to establish a dedicated UAS operator career field removes a major obstacle to efficient aircrew management, a great deal of turbulence will continue as

---

[30] News note from AirForce-magazine.com, February 21, 2008.

[31] This information is from AETC/A3R.

[32] Two authors of this monograph attended the Rated Management Conference with the responsible Project AIR FORCE Program Director. CSAF approval of the Air Force Operations, Plans and Requirements Conference minutes was confirmed by staff members at Headquarters Air Force and ACC. In a September 2008 message, however, Lt Gen Daniel J. Darnell (AF/A3/5) directed that UAS assignments would be an option for SUPT graduates starting in October 2008 and that "those graduates receiving UAS assignments will receive a follow-on assignment to a manned aircraft after their first tour in the UAS" (Darnell, 2008).

the Air Force struggles to meet a demand from combatant commanders (and a Secretary of Defense objective) to double the number of Predator and Reaper combat air patrols supporting global operations 24 hours a day, seven days a week (24/7), from 25 in June 2008 to 50 in FY 2011. The supervisors and instructors, for example, required to support the higher demand for UAS missions will continue to come from a pool of pilots who have crossed over from manned systems for quite some time. This will continue to limit the supply of qualified pilots who have developed the skills required to perform staff and supervisory responsibilities in their original weapon systems.[33]

Conclusions

The recurring theme in this chapter has been that the Air Force pilot-production goal of 1,100 SUPT graduates per year and the related goal of producing about 330 fighter pilots per year cannot be maintained without serious negative consequences for the training and readiness of fighter units. This problem is related to aircraft infrastructure, fighter-pilot-production levels, and experience definitions. Without changes, F-15C and F-16 units would experience overmanned conditions, inexperienced pilots would not fly enough to become experienced in a first flying tour, and second flying tours would be difficult to obtain.

The reductions in programmed fighter-pilot production announced in May 2007 eased the problems slightly and, coupled with the May 2007 policy decision to transfer pilots with limited experience to other career paths, delayed the problem onset. However, our analyses indicate that these changes are not sufficient to prevent serious problems from recurring in the operational units during and following FY 2013. Solving those problems requires further reductions in the number of pilots who enter IFF and continue to become fighter pilots.

Additionally, the Air Force is still unable to develop and sustain adequate inventories of appropriately qualified people to fill its rated and other operational requirements. An indication of the continuing

---

[33] These issues are currently under examination in follow-on PAF projects.

problem is that assignment managers at AFPC were not able to fill their "must-fill" rated billets for two consecutive assignment cycles (summer and fall 2008). These must-fill billets are determined by the Rated Staff Allocation Plan (RSAP) process and include only authorizations that are required to be 100 percent filled, such as API-1 and API-6 flying billets. They do not include rated staff billets that are normally filled at lower rates in the RSAP process. In the assignment match for spring 2009 (accomplished in November 2008), only 96 rated officers were released to fill rated staff billets—only about 7 percent of the 1,350 rated staff requisitions for that assignment cycle.[34]

---

[34] None of these are must-fill billets, and some have been vacant for some time. Some of them are being filled temporarily by contractors. The information in this paragraph was provided by AFPC/DPAO.

CHAPTER FIVE

The Potential Role of Total Force Integration Initiatives

We have long supported using TFI initiatives to ease Air Force absorption and manning issues (Thie et al., 2004; Taylor et al., 2000; and Taylor et al., 2002). The ANG and Air Force Reserve possess sizable inventories of highly experienced rated officers and fighter aircraft, both of which could contribute significantly to alleviating manning shortfalls and absorption chokepoints, respectively. Unfortunately, recent TFI initiatives have contributed very little toward these ends. This chapter examines the issues and underlying causes that have prevented these TFI initiatives from contributing. We will first examine the sustainability of existing paradigms for fighter units in the three total force components.

No Total Force Component Can Be Sustained with Existing Paradigms

Under current policy, almost every experienced pilot in all three components is initially absorbed and operationally developed in active units. We have already examined how existing active pilot inventory shortfalls generated policy decisions that stressed the training environments in active operational fighter units. We will now examine whether the dwindling active PMAIs will continue to support guard and reserve unit pilot requirements, beginning with estimates of future numbers of available active pilots.

The last chapter showed that the smaller active PMAI fighter inventories of the future will limit active fighter absorption capacity to fewer than 200 pilots per year to avoid severely damaging the training opportunities available in the active units. What kind of active fighter pilot inventory can these productions numbers sustain? The Air Force tracks historical total active rated service (TARS) information for rated officers, and a 10-year average for fighter pilots yields an expected value of 13.81 years of active service as a rated officer. This yields a steady-state inventory projection of 2,762 (13.81 × 200) fighter pilots on active duty to fill a requirement that will approach 4,200 pilots by the end of the FYDP, generating a steady-state shortfall of at least 1,400 fighter pilots.[1]

Air Force Reserve and Air National Guard Manning Issues

Historically, the Air Force Reserve Command (AFRC) and ANG have gotten virtually all their fighter pilots by hiring experienced pilots separating from active service. They train few new pilots.[2] Currently, enough pilots separate from active service each year to sustain the required

---

[1] Data provided by AF/A3O-AT. Note that the requirement for fighter pilots is lower than what was expected in FY 2001 (4,400; see Chapter Two). The 10-year TARS average includes FYs 1999–2008. This requirement for fighter pilots could be viewed as a lower bound because we have made no adjustments to include a pro rata share of the 600 pilot billets that do not require experience in a specific weapon system. If the calculated weighted average were used to determine this share, it would add another 175 pilots to the fighter requirement. In addition, these calculations make no attempt to include proposed new requirements for RC-12 pilots (as many as 67 RC-12 aircraft may be introduced by FY 2009, according to Clary, 2006), so the actual shortfall could easily be larger than this estimate. Initial plans appear to call for RC-12 pilots to come from the same pool as UAS pilots (mostly fighter pilots), so the new aircraft may initially increase the shortfall of fighter pilots.

[2] Indeed, from FYs 1999 through 2007, the guard and reserve graduated, on average, about one new pilot per operational squadron per year from the IFF course. This calculation is based on information from AETC/A3R and ACC/A5B. Historically, the ANG has trained only a few new pilots, and existing directives specifically require these pilots to fly well above the ANG RAP standards for their first two years because their flying skills are much more perishable than the typical ANG pilot. This limits this practice to considerably less than one pilot per unit per year.

AFRC and ANG inventory of approximately 1,460 pilots.[3] As available eligible active cohorts continue to dwindle, however, AFRC and ANG units will no longer be able to fill their requirements with fully trained pilots who are separating from active duty. At some point, AFRC and ANG units would have to begin training some of their own pilots.

We estimated the size of the looming AFRC-ANG shortfall as follows. The maximum time a pilot will remain in the total (i.e., active plus AFRC and ANG) inventory is 20 years.[4] Since an average of 13.81 years is spent in active service, only 6.19 years remain for AFRC or ANG service. Given an absorption of 200 pilots per year, the future sustainable AFRC/ANG pilot inventory will be 1,238 (6.19 × 200) pilots. We recognize that pilots who plan to affiliate with the guard or reserve could be more likely to separate from active service earlier than average, so this may not represent a true upper bound on the potential fighter pilot inventories for the nonactive components. Historically, however, the ANG or AFRC has not hired every separating fighter pilot, and many separating pilots may be unwilling to affiliate. Thus, our number likely does reflect a reasonable limit. The fact remains that

---

[3] This estimate is based on a requirement for 1,219 API-1 and API-6 billets to man flying units, counting both operational units and FTUs, plus at least 250 staff and individual mobility augmentee billets that are not directly generated by flying units. These estimates use force structure and manning information provided by AF/A3O-AT. We recognize that guard and reserve force structure numbers could drop sharply at some point as legacy fighter systems are phased out and will address this issue later in this chapter.

[4] The aircrew management system for all three components deals only with requirements and inventories for officers in grades O-5 and below; assignment flows for officers in higher grades are managed separately. Because of the gap between commissioning and entering SUPT, a TARS of 20 years actually requires close to 22 CYOS. O-5 authorizations in guard and reserve flying units are limited to squadron commanders, operations officers, and a few other very specialized billets, with fairly stringent constraints on officers serving over grade once they become eligible for retirement. Since the O-5 billets are typically used as grooming billets for promotions to O-6, people serving in them rarely remain in units as O-5s for extended periods. Typically, a pilot will either have been promoted to O-6 or have become eligible for retirement, or both, after 20 years.

For consistency, we should start the clock on measuring a useful career when pilots complete UPT. This will ensure they have at least 21 CYOS after the completion of 20 years rated service because UPT requires a full year of training. Many would have more than 21 CYOS because of delays in entering and/or completing UPT.

a need for guard and reserve units to become self-sustaining by train-ing their own new fighter pilots will require significant changes in their current paradigm, which relies heavily on high unit experience levels and very limited numbers of newly trained pilots.

A Theoretical Upper Bound for Potential AFRC and ANG Contributions

Ignoring, for the moment, the distinction between active pilots and those in AFRC and ANG, we estimated the size of the total pilot inventory that could be sustained by the total fighter PMAI. In Chap-ter Four, we estimated that active aircraft will be able to absorb about 200 pilots per year. Figure 2.1 showed that, by FY 2011, the AFRC/ANG inventory will be about two-thirds of the active inventory and that this proportion will remain fairly constant thereafter. If we assume that each of these aircraft could absorb as many new pilots as each active aircraft, the combined PMAI fighter inventory could absorb a maximum of 334 pilots per year (200 + 0.67 × 200).[5]

As mentioned earlier, the maximum time a pilot will remain in the total force (i.e., active plus reserve or guard) inventory is 20 years. Combining these numbers, we estimate that, at maximum, the total fighter PMAI could sustain a total fighter pilot inventory of 6,680 (20 × 334) pilots.

We must compare this with the total requirement for fighter pilots. As mentioned earlier, the requirement for active pilots approaches 4,200 by the end of the FYDP, and the guard and reserve combined will need a fighter pilot inventory of approximately 1,460 pilots. Even if we add another 250 pilots to account for unspecified requirements, demographic mismatches, and imperfect inventory management, we get a total requirement of under 6,000 fighter pilots—well within the theoretical maximum sustainable inventory of 6,680 pilots.

---

[5] This force structure projection does not include the additional cuts proposed to repair programming disconnects in the FY 2010 POM.

Obstacles to Realizing the Theoretical Upper Bound

To make use of AFRC and ANG aircraft to absorb active pilots, we must ensure that all assigned pilots can continue to fly at rates that will provide adequate training. Non–active-duty units currently fly at lower UTE rates than active units because their experience advantage allows it, and the active RAP training program was designed to require less experienced pilots to receive more training per month than more experienced ones. These UTE and RAP distinctions generate, in turn, significant manpower and other resource differences between active and nonactive units. These concerns led to an effort to reconfigure existing units that were segregated by component to associate pilots from different components and allow them to fly the same airframes.

Unit Associations

The Air Force has a long history of associating personnel from separate total force components to fly the same airframes. The traditional method combines the personnel in what are now known as *classic associate* units, where reserve component personnel operate (or maintain) equipment assigned to an active unit at an active location. These associations were initially formed in tanker and transport units, but they were also tested and adopted for fighters during the 1990s. This type of association can improve experience conditions in an active unit but provides no additional aircraft to absorb and develop new pilots. This led the Air Force to pursue a reverse association, now called *active-associate* units, where active personnel operate (or maintain) equipment assigned to a reserve component at a reserve component location. This concept was originally proposed at the second four-star rated summit in 1999 because it provides additional force structure for absorbing new pilots, thereby increasing absorption capacity. Unfortunately, this increase in absorption capacity has not yet been effectively realized.[6] Cultural concerns and resource constraints caused the components

---

[6] The only active-associate unit currently operating (at McEntire ANG Base, South Carolina) has never absorbed an active pilot directly from UPT and the F-16 FTU B-course.

to adopt a total force absorption program that instead moves limited-experience active component pilots to guard and reserve units. This program, however, has not appreciably improved absorption capacities within the total force.[7]

Both types of association share the problem of requiring independent chains of command for each component represented. This makes them less absorption-efficient in terms of the number of overhead and supervisory (API-6) billets required per absorbable airframe because the dual structure requires more experienced pilots than would be needed to support an integrated chain of command. While we believe the Air Force will eventually require the most efficient means available to sustain its three components, we recognize the need for an evolutionary process for the Air Force to address existing cultural issues.

Absorption Capacity Potential in Active-Associate Units

Our work has shown that a notional 24-PMAI active fighter squadron can successfully absorb between six and eight new B-course graduates per year and still maintain an acceptable training environment (see Taylor et al., 2002, Chs. Five, Six).[8] These numbers are approximate because a unit's absorption capacity depends on a number of parameters. These include its PMAI, UTE rates, ASDs, manning and experience levels, and the proportion of FAIPs among the new pilots. Policy decisions, such as experience objectives, also play a considerable role in determining unit absorption capacities. Indeed, keeping other parameters relatively constant, the low end of this absorption range roughly corresponds to an experience objective of 60 percent, and the high end to one of 50 percent.

It will be useful to frame this discussion in terms of absorption capacity per PMAI airframe to make meaningful comparisons among various types (and sizes) of units.[9] Eighteen-PMAI active squadrons,

---

[7]   See Taylor et al., 2002, pp. 92–95, and Taylor, 2004, pp. 37–39, for lengthier discussions of total force absorption program problems and issues.

[8]   The six-to-eight range applies specifically to F-16 and F-15C squadrons; A-10s and F-15Es can absorb slightly more.

[9]   This discussion is again based on Marken et al. 2007.

while less absorption-efficient than 24-PMAI squadrons, could still accept between 4.5 and 5.5 new pilots per year and maintain manning and experience levels that would provide effective training environments. These numbers yield an approximate absorption capacity per PMAI of 0.25 to 0.33 new pilots per PMAI for all active units. In 2003, we examined potential unit configurations for active-associate units that would yield reasonable absorption capacities (per airframe) for active pilots without requiring excessive resource expenditures. Our analyses indicated that a reasonable compromise could be achieved in a 24-PMAI active-associate unit flying current CAF-standard UTE rates, provided its guard or reserve pilot (API-1 and API-6) manning remained the same as the unit's end–FY 2006 manning authorization as a 15-PMAI unit. Such a unit would be able to absorb between four and five new pilots per year, which yields an active contingent of 14 to 16 pilots in the unit, using a nominal three-year tour and assuming inclusion of two previously experienced, IP-qualified, active pilots to provide the supervisory support that the Air Force requires for new pilots in an initial operational assignment. This unit configuration allows all pilots to fly at their respective programmed sortie rates.[10]

The original concept was to consolidate PMAI and O&M manpower authorizations from existing units to limit the increase in required resources to the amount needed to boost the UTE rates from the traditional guard and reserve values to the active programmed rates. This UTE increase is quite modest compared to the UTE rates required for these units to transition to a completely autonomous mode in which they absorb and develop all their own required pilots.

Smaller active-associate units are not nearly as absorption efficient. An 18-PMAI F-16 unit, manned at the original ANG 15-PMAI manpower authorizations, for example, would be able to absorb less than

---

[10] We used the following programmed sortie rates (in SCM) in a notional F-16 unit:
- AFRC/ANG: for part-time pilots, 6; for full-time pilots, 9.
- Active: for CMR pilots, 12; for BMC pilots, 7.

It should be noted that these are the programmed sortie rates that were developed when the RAP was initially implemented and should not be confused with the minimum sortie rates that are published in the RAP tasking documents.

one active pilot per year, even if it were able to increase its UTE rates to the active objective value of 18.4 sorties per airframe per month.[11] This yields an absorption capacity per PMAI of less than 0.042 pilots per airframe per year, which is less than one-fourth as absorption-efficient as a 24-PMAI unit.

The Air Force used this analysis in 2004 as it developed a proposal to reorganize guard and reserve F-16 units into 24-PMAI units as part of its 2005 Base Realignment and Closure program for 2005 (BRAC 2005). AFRC embraced this proposal and agreed to convert to two 24-PMAI units. ANG, however, rejected it and lobbied success-fully to realign primarily into 18-PMAI units instead of the 24-PMAI unit configuration that Headquarters Air Force had proposed. ANG's position was understandable: Giving up fighter aircraft to consoli-dated units would mean, at best, less training for ANG pilots and, at worst, loss of a flying mission.[12] Nonetheless, this cut the potential, fully resourced absorption capacity of guard and reserve F-16 units by a factor of almost four because the guard has over 85 percent of the post-BRAC reserve component F-16 PMAI. Further, the resources required to achieve the additional absorption that could be derived from active-associate units have never been provided. Thus, no active-associate fighter units have been formed.

Resource Issues for Active-Associate Units

The original concept for active-associate fighter units was to assign active-duty O&M personnel to existing guard and reserve units to fill out the crew ratios associated with increased PMAI authorizations and to operate the aircraft at sufficiently high UTE rates to train the assigned pilots at appropriate programmed RAP sortie rates.[13] This concept was

---

[11] This is because the unit overhead for ANG personnel is about the same for an 18-PMAI unit as for a 24-PMAI unit, leaving less room for active-component pilots.

[12] This is compelling evidence that preserving fighter units and the corresponding pilot bil-lets is important to ANG, if not to AFRC, and is clearly much less important to the active component. In the long term, though, the loss of total force absorption capability could mean that ANG units retain aircraft but have more difficulty finding pilots to fly them.

[13] These are the sortie rates listed in footnote 10.

based on the underlying assumption that these active personnel would become available from existing active units as they began to downsize and as additional classic associations were formed with the active units that remained.[14] The problem, though, was that active maintenance units were manned at only about 85 percent of their authorizations and were taking additional personnel cuts mandated by program budget decision 720 (PBD 720) that would reduce manning even further.[15] Thus, additional active maintenance manpower was unlikely to become readily available, and Air Force logisticians were unwilling to embrace any plan that would not ensure funding of adequate maintenance resources to produce the required sorties.

An independent RAND research study, conducted after BRAC 2005, determined that a notional 18-PMAI active-associate guard F-16 unit could boost its UTE rates from historical levels to the CAF standard of 18.4 sorties per airframe per month by converting existing part-time authorizations to full-time billets. The reason for this is the enhanced maintenance productivity generated in ANG units by their highly experienced, full-time guard maintainers, who are also less heavily tasked with ancillary duties than are their less-experienced active-duty counterparts. The total number of additional full-time billets required ranges from 32 to 45, depending on whether the unit

---

[14] Several factors influenced this line of thinking as the planning for the BRAC 2005 input evolved: (1) Classic associations were under development at Hill, Langley, and Elmendorf AFBs; associations were under consideration at Holloman AFB; and an expanded association was under examination for Shaw AFB/McEntire ANGB. (2) It was becoming quite clear to Air Force planners that the combined F-22 and F-35 buys would never approach the numbers of legacy fighters (i.e., A/OA-10s, F-15s, and F-16s) that they were replacing. (3) Serious pilot overmanning issues were projected for active operational fighter units.

[15] Secretary of Defense Michael Wynn testified to Congress that,

[w]ith Program Budget Decision (PBD) 720, the Air Force planned to reduce by 40,000 active duty, guard, reserve and civilian Full-time Equivalents (FTEs) to self-finance the critical recapitalization and modernization of our aircraft, missile and space inventories. (Wynn, 2007)

On June 11, 2008 the Secretary of Defense announced that these cuts would be discontinued (Hoffman, 2008).

retains a single-shift maintenance operation or converts to the two-shift standard normal for active units.[16]

The analysis cited in the preceding paragraph could point the Air Force toward a relatively straightforward means of providing the added maintenance required for an active-associate fighter unit to contribute the additional absorption capacity that is currently required, but the 18-PMAI/24-PMAI issue remains unresolved in almost all the ANG fighter units. An appreciable contribution to an increased absorption capacity still requires larger unit aircraft inventories.[17]

TFI initiatives must be fully resourced to be effective. These resource requirements include manpower authorizations, personnel allocations, support equipment, flying hours (including all required consumable commodities), aircraft parts, and several kinds of facilities, all in addition to PMAI. Most of these resource issues have not yet been addressed in existing TFI initiatives, and that is why no active-associate units are currently in operation. Indeed, as we note below, every association adapted to date as a TFI initiative has reduced, not increased, the overall absorption capacity of the total force.

What Needs to Be Done?

TFI is important to all three air components because it can help avoid the operating inefficiencies associated with three autonomous organizations having stovepiped and duplicated staff support requirements.

Current Paradigm Obsolete?

The TFI program could consider improvements to organizational structures and other means of enabling mission accomplishment and ensuring development and sustainment of required personnel invento-

---

[16] See Drew et al., 2008, Chapter Two, for the development of the increased maintenance productivity and Chapter Three, especially Table 3.2 and its supporting discussion, for the analysis that quantifies the full-time conversions required. Also, Chapter One quantifies the PBD 720 cuts.

[17] The political pressures imposed by (and on) governors and senators to ensure that ANG jobs and resources remain within their state boundaries may render this issue irresolvable.

ries in adequate numbers. The paradigm that governs all the existing organizational structures, operational methods, political motivations, and public laws may no longer be pertinent now that the Air Force is completing its transition to an Air Expeditionary Force and simultaneously conducts increased homeland defense activities to support counterterrorist operations. Post-9/11 operations and activities have already blurred distinctions between the ANG's Title 10 (support national emergencies) and Title 32 (organize, train, and equip) missions. For example, in air defense alert detachments that the ANG was tasked to establish to support homeland defense operations, a number of ANG personnel were recalled to full-time status. The Guard Bureau opted to place these full-time billets in Title 32 status to maintain command and control through their existing organizational structures because they viewed the positions as an "organize, train, and equip" responsibility. Their civilian employers are insisting that some of these individuals return to work or lose their original jobs because they have exceeded the five-year limit for military leave, even though the alert requirement remains valid and is clearly in support of a national emergency as part of Operation Noble Eagle.[18]

In the new paradigm, the total force must support extended operational deployments in various locations, while relying on only limited mobilization or even completely voluntary support from the reserve component. An increased cadre of deployable active-duty personnel associated with these units would seem to improve their ability to provide expeditionary support. The Air Force has been conducting continual combat operations in various locations since Operation Desert Storm began in January 1991. It may be prudent to accept that these conditions may continue indefinitely and restructure the total force accordingly. Yet, a recurring theme for all three components in many recent TFI initiatives has been to preserve, as much as possi-

---

[18] New legislation has been proposed to address this disparity because the original Public Law specifies only that "Title-10 service in support of a national emergency" is exempt from the five-year limit. It does not appear, however, that it will pass in time to help the majority of the affected personnel. The civilian employers that are insisting that the national emergency exemption does not apply include several major airlines, such as American Airlines. Every TFI initiative that the CAF has attempted has had Title 10/32 issues to overcome.

ble, the status quo for the affected units, thereby clinging to the old paradigm.

Three major initiatives—one each at Robins, Langley, and Hill AFBs—did not reduce guard or reserve aircrew authorizations, even though it meant decreasing active API-1 aircrew positions in the resulting units. Decreasing active aircrew positions always reduces a unit's absorption capacity. It would be far better, from an absorption perspective, to manage the mix of active and reserve billets in the resulting units to improve the overall absorption capacity per PMAI airframe. It would also be worthwhile to manage the mix of active, full-time reserve, and part-time reserve authorizations within the unit to improve its overall combat capability. It is not clear that this has been done in the existing TFI projects.[19]

Revised Directives

Existing policies that govern war mobilization processes and ongoing defense review efforts have imposed significant constraints that impede progress toward a new paradigm. For example, War Mobilization Planning Factor documents 3 and 5, which govern wartime crew ratios and other planning factors, need to be brought in line with the unit type code and readiness reporting requirements that are evolving as the Air Force completes its transformation into an air and space expeditionary force. The 2006 and 2008 Operational Availability scenarios, which govern the periodic BRAC and Quadrennial Defense Review processes, may also require revision to meet Air Force mission changes. Finally, the public laws that govern Title 10 and Title 32 distinctions, as well as those that address full-time personnel authorizations and uses, were all developed before the advent of the air and space expeditionary force concept.

---

[19] The unit at Robins, for example, was created following congressional action and a GAO investigation to overturn PBD 824, which the Secretary of Defense had issued in June 2001. A political compromise was reached to convert the ANG's 162 Bomb Wing to a "total force" air control wing with very little direction on how this should be achieved, leaving the ANG and ACC to put the pieces together. This comment is based on a "Lessons Learned" briefing presented by the 162 ACW Chief of Staff.

The Air Force might consider balancing billet authorizations across the entire spectrum of TFI initiatives, using additional classic associations, for example, to maintain an appropriate total number of reserve billet authorizations, rather than requiring a fixed number at each location. Also, additional active associations could be used to maintain the desired active billet totals, especially when existing TFI initiatives have already reduced the overall number of active API-1 authorizations, which remain the key element of the aircrew absorption and development process for all three components. Finally, flying units should be situated for effective and efficient operation and should receive adequate PMAI authorizations to be absorption-efficient, instead of remaining at existing locations with smaller PMAI authorizations. This issue will become even more important as legacy fighters continue to age with only limited procurement of replacement systems.[20]

Alleviating Current Rated Officer Shortfalls

Because of the pilot-production bathtub we discussed in Chapter Two, the Air Force currently has a shortage of field-grade rated officers. The shortage is more severe for officers with operational experience in CAF systems because of the absorption constraints that have plagued both fighter and bomber units since the initial four-star rated summit in 1996 restored aircrew production. These constraints did not exist in MAF units. SOF shortages also exist because of pipeline constraints and expanding force structure and mission requirements, and the force has many unfilled RSAP-controlled billets because its officers are filling disproportionate shares of must-fill and unspecified billets.

A large number of guard and reserve officers have the experience, backgrounds, and qualifications required to fill many of these billets. Many full-time ARC officers already serve in staff positions created through statutory tours and other means.[21] Policy decisions that would

---

[20] We recognize that UAS units will comprise a proportion of the replacements.

[21] Our primary concern here is that guard and reserve officers serving in staff billets should normally do so in full-time status. This could include Active Guard Reserve or technician status, or even an as-yet-to-be-developed capacity, in addition to extended active duty or other existing options.

enable qualified officers to fill empty active billets for two or three years could reduce the problem significantly. Additional policies that motivate qualified officers to voluntarily accept such tours would also be useful. In fact, the Air Force's Rated Force Policy Office conducted a survey of ARC pilots in August 2007 to gauge general interest in accepting active-duty staff positions and to determine the types of inducements or benefits that would encourage people to serve in such assignments.[22]

Summary

If the Air Force continues to assert a need for over 4,200 active-duty fighter pilots to fill increasing staff requirements (at the same time that cockpit requirements are decreasing), a total force approach that uses the fighter aircraft resources in the active and ARC inventories can support this need and the requirements for the ARC. Doing so, however, will require unit structure, maintenance, and personnel management changes. Support for these changes would depend not only on political and cultural considerations but also on their costs—which have yet to be determined.

Active-associate units are not as efficient at absorbing new pilots as active units are. If they could become more efficient, however, the combination of TFI initiatives and establishing a UAS career field that absorbs pilots and is considered a source for fighter staff positions could indeed eventually resolve CAF manning issues.

---

[22] Email, dated October 5, 2007, from the Chief, Rated Force Policy (AF/A1PPR). Preliminary results at that time, based on 564 responses (over 1,900 surveys were distributed), indicated that 44 percent were not interested in active-duty assignments; 11 percent were very interested; 21 percent were slightly interested; 13 percent were somewhat interested; and 11 percent were interested.

CHAPTER SIX

Conclusions and Recommendations

Decisions made to solve short-term aircrew management problems in one area can easily lead to long-term problems in another, as Chapter Two's brief summary of the changing fortunes of the fighter pilot community since the 1990s showed. Because the demand for experienced fighter pilots is increasing but the capability to produce them is decreasing, it is important to understand the short- and long-term implications of policy changes. Our dynamic model is a useful tool in this regard. It not only assists in the development of solutions to known problems but also reveals previously unanticipated effects of policy changes, such as the second-tour choke point.

Air Force circumstances have changed: Throughout the late 1990s and into 2000 and 2001, one primary reason that there were too few pilots was that the separation rates for pilots reaching the ends of their active-duty service commitments were high—induced in part by opportunities for jobs with commercial airlines. The beginning of the pilot bathtub aggravated the shortage, and the Air Force understandably hoped to reduce the shortage by increasing the production of pilots. Our modeling showed, however, that various constraints (aircraft inventory, available flying hours, and UTE rates) limited the number of fighter pilots who could be absorbed into the operational units and developed into experienced pilots. There are too few pilots to meet current requirements, but our analyses confirm that the ability of Air Force fighter units to absorb and develop new pilots continues to decrease. It is unlikely that these units will be able to produce fighter pilots at the rate required to satisfy the current demand for them. While we have argued that normalizing the UAS career field is a necessary condition to resolve the pilot inventory problem,[1] it is not sufficient to resolve these issues. In any case, a large number of pilots will be required to support these systems (as well as new developments, such as RC-12s[2]) as they transition to their own steady state, which will take at least a decade.

Our primary conclusion is that maintaining the health of fighter units requires reducing the number of new pilots entering them, ultimately to below 200 per year by 2016. Overabsorption decreases the number of monthly sorties inexperienced pilots can fly, lowers the average experience level of fighter units, makes it difficult or impossible for new pilots to become experienced in an initial three-year tour, and increases the amount of time a pilot must wait between a first flying tour and an opportunity for a second one. The last, in particular, slows the pilot's development of the background and experience elements needed to become useful in staff or supervisory positions. All these circumstances can lead to a general degradation of pilot skills and combat capability, as occurred at Pope AFB in 2000.

These conclusions are based on existing policies. Absorption capacity could be increased by increasing the number of absorbable cockpits, but the trend is in the opposite direction. Increasing UTE rates would also help increase absorption capacity, but legacy aircraft are aging rapidly, and every fighter MDS is currently struggling to attain CAF-standard UTE rates. Additionally, flying higher UTE rates would require additional funding for flying hours, which is unlikely.[3] Decreasing the number of flying hours required to become an experienced pilot would increase the absorption capacity of fighter units. However, as documented in our earlier work (Taylor, 2000; Taylor, 2002), these

---

[1]   That is, making the UAS a weapon system that is absorbable and one to which personnel can be assigned for a career track instead of just as a one-time ALFA tour.

[2]   The RC-12 is a modified Beechcraft King Air turboprop aircraft that "is part of the Air Force's stated goal of providing more manned and unmanned ISR capability inside theaters, especially Operations Iraqi Freedom and Enduring Freedom" (see LaGrone, 2008).

[3]   Historical UTE data are maintained by ACC's directorate of Maintenance and Logistics (ACC/A4P). ACC's Air and Space Operations, Flight Management Branch documents flying hour funding cuts for both the FY 2008 and FY 2010 POM cycles.

pilots would have less knowledge and be less capable and would contribute to the second-tour choke point described in Chapter Four.

While it is also possible to credit simulator hours to replace flying
hours in certain aspects of the training required to gain experience and
to develop new pilots, we do not see this as a viable near-term solution.
The added simulator training would require making adequate numbers
of high fidelity simulators available to handle the increase in simulator
hours; in addition, the operational flight profile software in the simulators would have to kept current with the latest version in the aircraft.
Both issues could require additional simulator system funding, which
would be difficult to obtain.[4] Even though simulator training may
become more important for fifth-generation systems because of issues
that limit the effectiveness of live training, the consequences for RAP
requirements and absorption capacity will require additional analysis.[5]

The options for increasing the absorption capacity of the active
fighter force are extremely limited, yet the need to decrease the flow of
fighter pilots through the system remains compelling. Simultaneously,
reducing the flow will reduce the inventory of fighter pilots available
for rated staff positions. A comprehensive solution will require the Air
Force to address both a demand problem and a supply problem. The
demand problem can be mitigated in at least two ways:

1. Reduce demand by closely examining emerging staff require-
    ments and eliminating those that are not genuinely required.

---

[4] The only dedicated funding program element for CAF simulators currently addresses
legacy systems only and would be unlikely to keep pace with the funding needed to exploit
the development of any future technologies and capabilities. Future simulator capabilities for
fifth-generation fighters must compete within the weapon system program element for funding and are unlikely to do so effectively when decisions are reduced to funding simulators
rather than aircraft or rather than increasing the combat capability of existing airframes by
enabling them to employ newly developing systems, such as the small diameter bomb.

[5] Taylor et al., 2000, addresses problems associated with trying to increase absorption by
lowering training standards. Marken et al., 2007, documents the essential elements for effective advanced simulator training. Also note that the F-22 still does not effectively absorb new
pilots out of UPT. The initial token test to make it an absorbable fighter platform was run
late in FY 2008.

2. For positions that are valid and that do appear to require personnel with fighter pilot experience, explore the full potential of other available manning alternatives, such as appropriately developed personnel currently affiliated with the ANG and reserve, CEAs, and civilians with prior military experience.

The supply problem can also be addressed in at least two ways:

1. Increase the supply of fighter pilots by effectively using the total force aircraft inventory (TFI: active, ANG, and Air Force Reserve force structure) to absorb and develop new active pilots.

2. Increase the supply of UAS pilots by establishing an independent, self-sustaining UAS career field. The current requirement that UAS operators who are fighter pilots must be able to return to a fighter unit is unmanageable—there are not enough fighter aircraft to allow it. Creating a UAS career field (and not treating it as an ALFA tour) will decrease stress on fighter units and increase the supply of UAS operators. A short-term solution here might involve sending SUPT graduates to UAS tours. A longer-term solution might involve changing the requirements for UAS operators (requiring, for example, only the first half of SUPT to become one).

It is critical that the Air Force curtail the flow of new pilots into active fighter units to avoid exceeding the current absorption constraints of the training system. In the short term, reducing this flow could lead to shortfalls for some Air Force needs, but the supply and demand options listed above will still allow the Air Force to meet expanding needs in other areas. Failing to reduce the flow will, in the near future, damage the combat capability of fighter units.

A Model for Dynamically Tracking Fighter Pilots Through Operational Squadrons

This appendix describes the current version (as of August 2007) of our dynamic model for tracking fighter pilots from the start of a first tour in an operational fighter squadron through the end of a second operational tour, in the context of the evolution of the fighter force structure over time and the policies for manning it. The model calculates the number of third-tour pilots operational squadrons require at each point in time—e.g., commanders, operations officers, wing weapons officers, and standards and evaluation officers. However, because it does not track pilots' careers through the third tour, the model says nothing about the inventory of candidates for these positions.

The model is written in the GAMS language.[1] GAMS was designed and is generally used to generate the large data structures used in mathematical programming (i.e., optimization) models, and the GAMS application is distributed with a number of powerful solvers for such problems. Our model, however, does not involve optimization. Instead, we take advantage of GAMS' array manipulation capabilities to implement our model as a deterministic simulation.[2]

The model steps through time in intervals of one month. In each month, it iteratively adjusts the number of hours each inexperienced

---

[1] See the GAMS Web site for more information.

[2] If that term sounds like an oxymoron, it is because one often assumes that a simulation model must have random elements, in which case it is often called a Monte Carlo simulation model. Our model is a simulation with no random elements, which is why we call it a *deterministic simulation*.

pilot flies to bring the demand for inexperienced flying hours into balance with the supply. This appendix begins with calculations of the demand for and supply of flying hours for inexperienced pilots, then describes how we converge on the correct number of hours flown each month per inexperienced pilot.

The history of the model's development (see Chapter Three) makes clear the reasons for many of the model's features that might otherwise appear to be confusing and unnecessarily complicated.

The Longitudinal and Cross-Sectional Points of View

The model examines the inventory of fighter pilots from two points of view. In the *longitudinal* point of view, the model follows each cohort of pilots through their careers, from the time they start a first tour in an operational fighter squadron to the time they exit a second operational tour (or the end of the simulation, if that comes first). Each pilot passes through the same sequence of stages:

1. inexperienced pilot in a first operational tour
2. experienced pilot in a first operational tour
3. planned gap between first and second tours (generally an ALFA tour)
4. unplanned delay while waiting for a second-tour billet to become available
5. second operational tour.

Different pilots may spend different amounts of time in any one of these stages.

In the *cross-sectional* point of view, the model examines the entire inventory of pilots at a single point in time. This inventory will include pilots from different cohorts and at different stages in their careers.

The two points of view interact through the mechanism of the fighter force structure and the policies for manning it. At any time, a limited number of billets is available for first- and second-tour pilots, a specified number of which must be occupied by experienced pilots. The aircraft in the force structure can generate only a limited number of flying hours, and the number that can be allocated to inexperienced pilots is strictly constrained. (Recall that inexperienced pilots become experienced by accumulating flying hours.) The model reconciles the two points of view by adjusting the amounts of time each cohort spends in each of the five stages.

The Longitudinal Equations

Total Pilots in a First Operational Tour (Stages 1 and 2)

The model currently contains two types of pilots entering a first operational tour: pilots who enter the B-course directly from UPT (*ptype = UPT*) and pilots who enter the B-course after completing an initial tour as an IP in trainers (*ptype = FAIP*). In the model, most pilots serve a first operational tour in an active-duty fighter squadron, but we have made provision for limited numbers of pilots to serve the first tour in a reserve-component squadron.[3] The following arrays include all first-tour pilots, wherever they serve a first tour:

Entry1 (*mo, ptype*) = Number of pilots of type *ptype* entering a first operational tour in month *mo*

Tour1 (*mo, ptype, tos*) = Number of first-tour pilots of type *ptype* who, as of month *mo*, have been on station *tos* months.

We assume that all pilots of the same type who enter during the same month and are allowed to complete their tours will exit their tours in the same month. So, let

T1len (*mo, ptype*) = Length of a first tour for pilots of type *ptype* as of month *mo*

Exit1 (*mo, ptype*) = Number of pilots of type *ptype* exiting a first operational tour in month *mo*.

---

[3]   The reserve component encompasses Air Force Reserve and ANG squadrons.

As mentioned in Chapter Three, the user can specify a number of first-tour pilots to depart early, before they complete their tours:

DepEarly1 (*mo, ptype*) = Number of pilots of type *ptype* who depart early, in month *mo*.

It is also possible to identify the times during a tour when the pilot may depart early (and to make specific adjustments):

TosEarly1 (*tos*) = Relative weight given to departing early during month *tos* on station

MltEarly1 (*mo, ptype*) = Adjustable multiplier for making early departures match specified numbers.

If *TosEarly1* (*mo*) = 0, then no pilot in month *mo* of the tour will depart. If *TosEarly1* (*mo*) > 0, then the model allows a pilot in month *mo* of the tour to depart early. Increasing the value of *TosEarly1* (*mo*) makes it more likely that a pilot will depart in month *mo*. To date, we have assumed that no FAIPs depart early, i.e., that *DepEarly1* (*mo, FAIP*) = 0.

Given the entry rates and the tour lengths,

If *tos* = 1, then

$$Temp1\left(mo, ptype, tos\right) = Entry1\left(mo, ptype\right).$$

If $1 < tos \leq T1len\left(mo, ptype\right)$, then

$$Temp1\left(mo, ptype, tos\right) = Tour1\left(mo - 1, ptype, tos - 1\right).$$

If $1 < tos \leq T1len\left(mo, ptype\right)$, then

$$Temp1\left(mo, ptype, tos\right) = 0.$$

The array *Temp1* (*mo, ptype, tos*) contains the pilots who would be in their first tour if no pilots depart early. The equation adds new pilots at the start of their tours, that is, if *tos* = 1; moves pilots one month deeper into the first tour for each month of calendar time that passes, that is, if $1 <$ *tos* $\leq$ *T1len* (*mo, ptype*); and drops pilots out of the system when they reach the end of their tours, that is, if *T1len* (*mo, ptype*) < *tos*.

The array *OutEarly1* (*mo, ptype, tos*) contains the number of pilots that depart their first tours early, i.e., by *tos* = *T1len* (*mo, ptype*):

$$OutEarly1\left(mo,\ ptype,tos\right) = \left[\frac{MltEarly1\left(mo,\ ptype\right) \times TosEarly1\left(tos\right)}{1 + MltEarly1\left(mo,\ ptype\right) \times TosEarly1\left(tos\right)}\right]$$
$$\times\ Temp1\left(mo,\ pype,tos\right).$$

The multiplier *MltEarly1* (*mo, ptype*) must be adjusted so that

$$DepEarly1\left(mo,\ ptype\right) = \sum_{tos} OutEarly1\left(mo,\ ptype,tos\right).$$

We do this with an iterative binary search algorithm. At each iteration, the multiplier is taken to be somewhere in the interval between a lower bound (initialized to zero before iteration 1) and an upper bound (initialized to a very large number before iteration 1). The model sets the multiplier equal to the midpoint between the lower and upper bounds and tests whether that multiplier causes too many or too few pilots to depart early. If too many depart, the next iteration takes the multiplier to be in the lower half of the previous interval. If too few depart, the next iteration takes the multiplier to be in the upper half. Thus, at each iteration, the interval containing the correct value of the multiplier decreases by half.

The multiplier converges to zero if *DepEarly1* (*mo, ptype*) = 0, and to the very large upper bound if *DepEarly1* (*mo, ptype*) is larger than the number of pilots in the time-on-station window for early departure.[4] In this case, all the pilots in the window depart early. If fewer pilots depart

---

4   The window consists of all values of *tos* for which *TosEarly1* (*tos*) > 0. The equations given are only correct as long as the largest *tos* in the window is smaller than the length of the first tour, *T1len* (*mo, ptype*).

early than are in the window, they are taken preferentially from values of tos for which $TosEarly1$ (*tos*) is largest.

Next, we calculate

$$Tour1\left(mo, ptype, tos\right) = \left[\frac{1}{1 + MltEarly1\left(mo, ptype\right) \times TosEarly1\left(tos\right)}\right]$$

$$\times Temp1\left(mo, pype, tos\right). \qquad \text{(A.1)}$$

A quick check will show that

$$Tour1\ (mo, ptype, tos) = Temp1\ (mo, ptype, tos) - OutEarly\ (mo, ptype, tos).$$

That is, the number of pilots remaining in the system equals the number of pilots who would remain if none left early less the number who do leave early.

Clearly, the number of first-tour pilots of type *ptype* at month *mo* must equal the number at month *mo*–1 plus gains minus losses. The term $Exit1$ (*mo, ptype*) is one of the loss terms. The other is the sum over *tos* of $OutEarly1$ (*mo, ptype, tos*). The term $Entry1$ (*mo, ptype*) represents gains:

$$Exit1\left(mo, ptype\right) = \sum_{tos=1}^{T1len\left(mo-1, ptype\right)} Tour1\left(mo-1, ptype, tos\right) + Entry1\left(mo, ptype\right)$$

$$- \sum_{tos=1}^{T1len\left(mo, ptype\right)} Tour1\left(mo, ptype, tos\right)$$

$$- \sum_{tos=1}^{T1len\left(mo, ptype\right)} OutEarly1\left(mo, ptype, tos\right).$$

$$\text{(A.2)}$$

The sums of $Tour1$ (*mo*–1, *ptype, tos*) and $Tour1$ (*mo, ptype, tos*) are the numbers of pilots at months *mo–1* and *mo*, respectively.

In most versions of the model, *Entry1* (*mo, ptype*) is an input,[5] but some versions calculate the entry rates that will achieve a specified goal (e.g., the largest entry rate that will maintain a specified minimum experience level). Likewise, *T1len* (*mo, ptype*) is always at least 32 months. But the model will extend the tour length to as much as 36 months to allow pilots enough time to become experienced. In all versions, *Tour1* (*mo, ptype, tos*) is an input for the initial month (*mo* = 0, nominally the last month of FY 1999) and is calculated for all subsequent months. *Exit1* (*mo, ptype*) is always calculated.

Inexperienced Pilots in a First Operational Tour (Stage 1)

A fighter pilot is inexperienced until he has accumulated a specified number of flying hours in his primary mission aircraft. A pilot will accumulate most of these flying hours during his first tour in an operational fighter squadron.[6] Let

FH2E (*mo, ptype*) = The number of flying hours a pilot of type *ptype* must accumulate during his first operational tour to be deemed experienced.

For all months through FY 2006, *FH2E* (*mo, UPT*) = 420 flying hours (430 for the F-15C) and *FH2E* (*mo, FAIP*) = 220 flying hours (230 for the F-15C). In some model runs, however, we have changed these numbers as one way to represent a policy of allowing simulator hours to substitute for some flying hours.

We also introduce variables for the number of hours flown by each inexperienced pilot in each month. We assume that all inexperienced pilots fly the same amount in any given month, whether the first tour is in an active-duty or a reserve-component squadron. Of course, pilots may fly different amounts in different months. Later, we will describe how we adjust these quantities to achieve a balance between the demand for and supply of inexperienced flying hours.

---

[5] In practice, we allowed users to specify annual numbers of entries and assumed one-twelfth of the annual number enter in each month.

[6] The first 70 hours (for the F-15C) or 80 hours (for the A/OA-10, F-15E, and F-16) are accumulated during the B-course for the aircraft, which the pilot takes just prior to his first assignment to an operational fighter squadron.

In our model, only pilots in their first operational tour can be inexperienced, and they will be the pilots that have been on station too little time to have accumulated $FH2E$ (mo, $ptype$) flying hours. At any month mo, we count back, accumulating the hours inexperienced pilots have flown in previous months until we reach a month in which the accumulated number of hours finally equals or exceeds $FH2E$ (mo, $ptype$). Pilots who started a tour before then must be experienced. Pilots who started after that must be inexperienced. Pilots who started at that time may or may not be experienced. Let

HCM (mo) = Hours flown per inexperienced pilot in month mo.

If $\sum\limits_{j=1}^{tos} HCM\left(mo - j + 1\right) < FH2E\left(mo,\, ptype\right)$, then

$$Wt\left(mo,\, ptype,\, tos\right) = 1.$$

If $\sum\limits_{j=1}^{tos-1} HCM\left(mo - j + 1\right) < FH2E\left(mo,\, ptype\right)$ and

if $\sum\limits_{j=1}^{tos} HCM\left(mo - j + 1\right) \geq FH2E\left(mo,\, ptype\right)$, then

$$Wt\left(mo,\, ptype,\, tos\right) = \frac{FH2E\left(mo,\, ptype\right) - \sum\limits_{j=1}^{tos-1} HCM\left(mo - j + 1\right)}{HCM\left(mo - tos + 1\right)}.$$

If $\sum\limits_{j=1}^{tos-1} HCM\left(mo - j + 1\right) \geq FH2E\left(mo,\, ptype\right)$, then

$$Wt\left(mo,\, ptype,\, tos\right) = 0.$$

(A.3)

One can interpret $Wt\,(mo,\,ptype,\,tos)$ as the fraction of pilots in their tosth month on station who are not yet experienced. For small values of tos, the weight is one; for large values, the weight is zero. The weight can be a fraction for only one value of tos, the month during which the pilots actually fly hour number $FH2E\,(mo,\,ptype)$. Thus, the number of inexperienced pilots of type $ptype$ is

$$InexPlt\left(mo,\,ptype\right) = \sum_{tos=1}^{tosmax} Wt\left(mo,\,ptype,\,tos\right)$$
$$\times\ Tour1\left(mo,\,ptype,\,tos\right). \tag{A.4}$$

Note that the summation in Equation A.4 extends to a maximum of only $tosmax$ months on station. In the model, a pilot's first operational tour nominally lasts 32 months ($tosmin$ = 32), but the model will extend the tour by as much as four additional months (to 36 months) if necessary to allow pilots to become experienced. If pilots would not become experienced even after $tosmax$ = 36 months, they exit the operational squadron as inexperienced pilots.

The next few quantities are calculated and displayed as part of the output used for analysis. They are not used in subsequent calculations. The TTE, if pilots become experienced, is

$$TTE\left(mo,\,ptype\right) = \sum_{tos=1}^{tosmax} Wt\left(mo,\,ptype,\,tos\right). \tag{A.5}$$

Pilots can finish a first tour without becoming experienced only if $Wt\,(mo,\,ptype,\,tosmax)$ = 1. In this case, $TTE\,(mo,\,ptype)$ = $tosmax$. To determine the number of pilots who fail to become experienced,

If $Wt\left(mo,\,ptype,\,tosmax\right) < 1$, then

$$NeverExp\left(mo+1,\,ptype\right) = 0.$$

If $Wt\left(mo,\ ptype,\ tosmax\right)=1$, then

$$NeverExp\left(mo+1,\ ptype\right)=Tour1\left(mo,\ ptype,\ tosmax\right).$$

$$(A.6)$$

The number of pilots absorbed (i.e., who become experienced) in month *mo*, is

$$Absorb\left(mo,\ ptype\right)=InexPlt\left(mo-1,\ ptype\right)-InexPlt(mo,\ ptype)$$
$$+Entry1(mo,\ ptype)-NeverExp\left(mo,\ ptype\right).$$

$$(A.7)$$

Experienced Pilots in a First Operational Tour (Stage 2)

We calculate the length of the first tour as follows (the *CEIL* function equals the smallest integer at least as large as its argument):

$$T1len\left(mo,\ ptype\right)=MAX\left\{tosmin, CEIL\left[TTE\left(mo,\ ptype\right)\right]\right\}.\qquad(A.8)$$

The total number of first-tour pilots present in month *mo*, is

$$T1tot\left(mo,\ ptype\right)=\sum_{tos=1}^{T1len\left(mo,\ ptype\right)}Tour1\left(mo,\ ptype,\ tos\right).\qquad(A.9)$$

The number of experienced first-tour pilots, of course, is the difference between total first-tour pilots (Equation A.9) and inexperienced first-tour pilots (Equation A.4).

The above formulas for month *mo* all depend on values being known for *Tour1* (*mo–1, ptype, tos*) for all values of *pytpe* and *tos* and for *HCM* (*mo–j*) as far back in the past as *j* = *tosmax–1*. We have selected the year FY 2000 as the first year of our simulation. To get the process started, we initialize *Tour1* (*mo* = 0, *ptype, tos*) to equal the average monthly entry rate of type *ptype* pilots during or just before FY 2000 (our choice varies somewhat for different MDSs). We assume that *HCM* (*mo*) is the same for all months prior to FY 2000 and that *HCM* (*mo* = 0). The model thus calculates the steady-state number of inexperienced pilots and the steady-state amount of flying they do corresponding to the initial inventory *Tour1* (*mo* = 0, *ptype, tos*).[7]

Planned Gap Between First and Second Tours (Stage 3)

Only a few fighter pilots go straight to a second operational tour immediately after their first tours (ops-to-ops). Most pilots spend the two to four years after a first operational tour as an SUPT instructor or an air liaison officer—tours outside the pilots' primary mission aircraft and sometimes even nonflying—before receiving a second operational tour.

The model assumes that 10 percent of both pilot types (UPT and FAIP) go ops-to-ops, with a planned gap of six months between tours. For the remainder, there is a planned gap of 36 months between the first and second operational tours.

The model does not use the number of pilots between tours in any of its calculations, so there is no need to present an equation for this quantity.

Unplanned Delay While Waiting for a Second-Tour Billet (Stage 4)

An individual completing stage 3 enters a pool of pilots awaiting a second operational tour. We make no distinction among pilots in the pool. It makes no difference whether they are going ops-to-ops or have had an intervening tour. It also does not matter whether their *ptype* was UPT or FAIP.

As described earlier, 10 percent of both UPT and FAIP pilots enter the pool six months after completing a first tour. The remaining 90 percent of UPT pilots enter the pool 36 months after they complete a first tour, but the remaining 90 percent of FAIPs split into two groups. FAIPs that do not go ops-to-ops complete an active-duty service commitment at about the time they would have become eligible for a second tour; we assume that 50 percent of these FAIPs elect to

---

[7] The reader may wonder why we begin the simulation in FY 2000, rather than a year closer to the present, such as FY 2004 or FY 2005. By starting so early, we hope to reduce the influence of our *de facto* steady state assumption on the results for years of interest.

leave the active Air Force at this time. The remaining 40 percent of FAIPs enter the pool. Let

fdelay (*t, ptype*) = The fraction of type *ptype* pilots with a delay of *t* months between exiting a first operational tour and becoming eligible for a second operational tour.

The number of pilots entering the pool in month *mo*, then, is

$$EntPool\left(mo\right) = \sum_{ptype} \sum_{t} fdelay\left(t, ptype\right) \times Exit1\left(mo - t, ptype\right). \quad (A.10)$$

Pilots leave the pool to enter a second tour in their primary mission aircraft. These tours usually take place in operational squadrons but may take place in training units. If some pilots receive a first operational tour in a reserve-component squadron, some pilots in a second or subsequent tour will accompany them. Let

Entry2 (*mo*) = The number of pilots entering a second tour in month *mo*.

Then, the number of pilots in the pool at month *mo* is

$$Pool\left(mo\right) = Pool\left(mo - 1\right) + EntPool\left(mo\right) - Entry2\left(mo\right). \quad (A.11)$$

We assume that the pool is empty at the start of the simulation (the last month of FY 1999).

The next few quantities are calculated and displayed as part of the output used for analysis. They are not used in subsequent calculations. We assume that pilots are drawn from the pool on a first-in, first-out basis. We can then calculate the amount of time a pilot just entering the pool at the end of month *mo* will wait for his second-tour assignment:

$$WillWait\left(mo\right) = K + \alpha, \quad (A.12)$$

where K and α satisfy $0 \le \alpha < 1$ and

$$Pool\left(mo\right) = \sum_{t=1}^{K} Entry2\left(mo + t\right) + \alpha \times Entry2\left(mo + K + 1\right),$$

where K is the number of whole months of entries into a second tour (which are the same as exits from the pool) that stand between the newest entrant to the pool and his second-tour assignment and α is the remaining fraction of a month.

A similar calculation gives the time that a pilot assigned to his second tour at the end of month mo had to wait for it:

$$DidWait\left(mo\right) = L + \beta, \tag{A.13}$$

where L and β satisfy $0 \le \beta < 1$ and

$$Pool\left(mo\right) = \sum_{t=0}^{L-1} EntPool\left(mo - t\right) + \beta \times EntPool\left(mo - L\right),$$

where L is the number of whole months of entries into the pool that stand between the pilot just entering the second-tour assignment and entry into the pool and β is the remaining fraction of a month.

Second Operational Tour (Stage 5)

The equations for calculating the number of pilots in a second tour hardly differ from the equations for the first tour. Let

$Tour2$ (mo, tos) = The number of second-tour pilots who, as of month mo, have been on station tos months

$T2len$ = The length of the second tour

$Exit2$ (mo) = The number of pilots exiting a second operational tour in month mo.

In the model, most pilots will serve a second tour in their primary mission aircraft in an active-duty fighter squadron, but if some first-tour pilots are assigned to reserve-component squadrons, some pilots in a second or subsequent tour will be assigned to the reserve-component squadron as well. There may also be opportunities to serve a second tour as an instructor in an FTU. The arrays $Entry2$ (mo) and $Tour2$ (mo, tos) include all second-tour pilots, wherever they serve the second tour.

As mentioned in Chapter Three, the user can specify a number of second-tour pilots to depart early, before they complete their tours. Let

$DepEarly2\,(mo)$ = The number of second-tour pilots who depart early in month mo

$TosEarly2\,(tos)$ = The relative weight given to departing early during month tos on station

$MltEarly2\,(mo)$ = An adjustable multiplier for making early departures match specified numbers.

The model then calculates

If $tos = 1$, then

$$Temp2\big(mo, tos\big) = Entry2\big(mo\big).$$

If $1 < tos \le T2len$, then

$$Temp2\big(mo, tos\big) = Tour2\big(mo - 1, tos - 1\big).$$

If $T2len < tos$, then

$$Temp2\big(mo, tos\big) = 0.$$

and

$$OutEarly2\big(mo, tos\big) = \left[\frac{MltEarly2\big(mo\big) \times TosEarly2\big(tos\big)}{1 + MltEarly2\big(mo\big) \times TosEarly2\big(tos\big)} \right]$$
$$\times Temp2\big(mo, tos\big).$$

The multiplier $MltEarly2$ (mo, ptype) must be adjusted so that:

$$DepEarly2\big(mo\big) = \sum_{tos} OutEarly2\big(mo, tos\big).$$

As with first-tour early departures, we do this with a binary search algorithm.

Next, we calculate the following:

$$Tour2(mo, tos) = \left[\frac{1}{1 + MltEarly2(mo) \times TosEarly2(tos)} \right]$$

$$\times Temp2(mo, tos) \tag{A.14}$$

and

$$Exit2(mo) = \sum_{tos=1}^{T2len} Tour2(mo - 1, tos)$$

$$- \sum_{tos=1}^{T2len} Temp2(mo, tos) + Entry2(mo). \tag{A.15}$$

The total number of pilots in a second tour is

$$T2tot(mo) = \sum_{tos=1}^{T2len} Tour2(mo, tos). \tag{A.16}$$

In all versions of the model to date, $T2len$ is an input (nominally 32 months), while the other quantities—$Entry2(mo)$, $Tour2(mo, tos)$, and $Exit2(mo)$—are calculated.

Cross-Sectional Constraints on Available Billets

The two constraints on the inventory of pilots in each month derive from the need to man the operational squadrons. One requires the number of pilots to at least equal an authorized number. The other requires that a minimum number of the pilots be experienced.

The Force Structure

We start with the fighter force structure, including the historical force structure through FY 2005, and the projected future force structure. The force structure lists the fighter squadrons at the end of each future year by component and installation and gives the number of aircraft each squadron is authorized:

PAA (*mo, sq*) = The number of aircraft authorized for squadron *sq* in month *mo*.

The force structure specifies the authorized aircraft in active, guard, and reserve squadrons at the end of each fiscal year. We assume that a squadron has a full complement of aircraft at the start of the last year it appears in the force structure and these aircraft disappear from the squadron linearly over the course of that year. That is, if the squadron has 24 aircraft at the start of the year, it will have 22 aircraft one month into its last year, 20 aircraft two months into the year, and so on. Similarly, we assume that a squadron is full at the end of the first year it is introduced and that its inventory had increased linearly over the course of that year. We assume that squadrons present in the initial year of the simulation were present the year before, so they begin the year full.

Each squadron is either an active-duty or a reserve-component unit. We denote this by $sq \in AD$ or $sq \in RC$, respectively.

Authorized Manning of Active Squadrons

Authorizations by Aircrew Position Indicator (API). The Air Force describes a fighter squadron's authorized manning in terms of API. Only two are relevant for our purposes: API-1 and API-6.

API-1 pilots are assigned to the squadron, and their job is to fly the squadron's aircraft. The Air Force calculates a unit's authorization for API-1 pilots as the product of a crew ratio and the number of aircraft assigned to the unit, rounded up to the next integer. For the F-15C, F-15E, and F-16 aircraft, the crew ratio is 1.25 pilots per aircraft, so each 24-PAA squadron is authorized 30 (1.25 × 24) API-1 pilots, and each 18-PAA squadron is authorized 23 (1.25 × 18, rounded up) API-1 pilots.[8] The A/OA-10 currently has a crew ratio of 1.5. Before June 2005, A/OA-10 squadrons had a "split" crew ratio. Aircraft designated A-10 had a crew ratio of 1.5, while OA-10 aircraft had a crew ratio of 2.

---

[8]   Crew ratios can be found in AFI 65-503, 1994, Table A36-1, "Authorized Aircrew Composition—Active Forces."

The recipe for API-6 authorizations is not so simple, but we can approximate them as follows. Each squadron has a commander and an operations officer. Each wing has a standards and evaluation officer and a chief of weapons, shared among the squadrons in the wing. Each squadron has an additional 0.23 API-6 pilots per aircraft, again calculated one squadron at a time, rounded up, and summed over squadrons. We chose the factor 0.23 to give a 24-PAA squadron six additional API-6 pilots and an 18-PAA squadron five additional API-6 pilots. These rules generate authorizations for the FY 2005 force structure that closely match data obtained from a variety of sources.[9]

Combat Mission Ready Versus Basic Mission Capable. Once a pilot has been assigned or attached to a unit, the squadron commander may place him in either a CMR billet or a BMC billet. To be CMR, a pilot must be qualified and proficient in all the primary missions tasked to the assigned unit and weapon system. To be BMC, a pilot need only be familiar with all the primary missions tasked to the assigned unit and weapon system and may be qualified and proficient in some.[10]

RAP specifies how much a pilot must fly according to whether the pilot is assigned to a CMR or a BMC position and whether the pilot is experienced. In an F-15C unit in the active Air Force, for example, to maintain BMC status, a pilot must fly 72 sorties per year if inexperienced and 60 if experienced. To maintain CMR status, a pilot must fly 110 (inexperienced) or 98 (experienced) sorties per year. The pilot who fails to fly this much may be placed on probation for a month and, if unable to fly a month's worth of the required sorties during that time, regresses to non-BMC or non-CMR status.

Representing Authorized Manning of Active-Duty Squadrons in the Model. In the model, only the distinction between CMR and

---

[9] One source is an old spreadsheet that purports to be the Air Force's RAP model. The file, named "RAPFHNew (2).xls," is dated February 22, 2002. Every unit in this file, whatever the MDS and whether it has 18 or 24 aircraft, is assumed to have nine API-6 pilots. A second, more recent source is a spreadsheet from ACC, "CAF Manning Feb–May 05.xls," which we obtained in early June 2005. It suggests that API-6 authorizations are slightly less, perhaps eight for a 24-PAA squadron and as few as seven for an 18-PAA squadron.

[10] Paraphrased from AFI 11-2F15V1. There is a similar AFI for each MDS (AFI 11-2F15EV1, AFI 11-2A-OA-10V1, AFI 11-2F-16V1). Each contains a similar definition.

BMC is important because, as we will see later, it influences how many flying hours are allocated to inexperienced pilots. Let

$CR\ (mo,\ CMR)$ = Crew ratio for computing CMR authorized manning

$CR\ (mo,\ BMC)$ = Crew ratio for computing BMC authorized manning

$CMR6\ (mo)$ = Authorized API-6 pilots who fly at CMR rates, not calculated using crew ratios; includes squadron commander, operations officer, standards and evaluation officer, and weapon chief.

As described earlier, authorized API-1 manning is calculated as

$$CMR1\big(mo\big) = \sum_{sq \in AD} CEIL\Big[CR\big(mo, CMR\big) \times PAA\big(mo, sq\big)\Big]. \qquad \text{(A.17)}$$

Since all API-1 pilots fly at CMR rates, we use the notation $CMR1\ (mo)$. Similarly, authorized API-6 pilots who fly at BMC rates is calculated as

$$BMC6\big(mo\big) = \sum_{sq \in AD} CEIL\Big[CR\big(mo, BMC\big) \times PAA\big(mo, sq\big)\Big]. \qquad \text{(A.18)}$$

Authorized Active Pilots in Reserve-Component Squadrons

The Air Force does not generally use reserve-component aircraft for training active-duty pilots, and there is no official method for calculating the number of active-duty pilots involved. However, the Air Force has had difficulty absorbing an adequate number of pilots using active squadrons alone, and there have been discussions about and even tests of using reserve-component squadrons to increase the absorption capacity. In the model, we consider two cases. An 18-PAA reserve-component squadron fences off 15 aircraft for the use of reserve-component pilots, leaving three aircraft for active-duty pilots. We place four active-duty pilots in this squadron: three in a first operational tour and one in a second or subsequent operational tour. A 24-PAA reserve-component squadron also fences off 15 aircraft for the use of reserve-component pilots, leaving nine aircraft for active-duty pilots. We place

12 active-duty pilots in this squadron: ten in a first operational tour and two in a second or subsequent operational tour.

To turn off this feature, we fence off all reserve-component aircraft for use by reserve-component pilots, leaving none for use by active-duty pilots. In this case, we place no active-duty pilots in reserve-component squadrons. In the model,

AssocT1mx (*mo*) = Active first-tour pilots authorized for reserve-component squadrons

AssocT2mx (*mo*) = Active second- or subsequent-tour pilots authorized for reserve-component squadrons.

Authorized Active Second-Tour Pilots in Alternative Billets

There are billets other than those in an operational fighter squadron that provide pilots an opportunity to fly their primary mission aircraft, giving some an alternative to a second operational tour. The most common alternative billet is that of instructor in the primary aircraft's FTU, but there are a few others (e.g., pilot in an aggressor squadron). Only a fraction of such billets can be filled by pilots in a second tour; some must be filled by pilots who have already completed a second tour. In the model,

AuthAlt2 (*mo*) = The number of alternative second-tour billets.

Actual Manning by Active-Duty Pilots

First-Tour Pilots. Pilots in a first tour can be assigned to either an active-duty or a reserve-component squadron. We assume that the fraction of first-tour pilots that are inexperienced is the same in active-duty and reserve-component squadrons. Thus, if *AssocT1N* (*mo*) and *AssocT1X* (*mo*) are the numbers of inexperienced and experienced first-tour pilots, respectively, in reserve-component squadrons,

$$AssocT1N\left(mo\right) = \frac{\sum\limits_{ptype} InexPlt\left(mo, ptype\right)}{\sum\limits_{ptype} T1tot\left(mo, ptype\right)} \times MIN\left\{\begin{array}{l} AssocT1mx\left(mo\right) \\ \sum\limits_{ptype} T1tot\left(mo, ptype\right) \end{array}\right\} \quad (A.19)$$

and

$$AssocT1X\left(mo\right) = MIN \left\{ \begin{array}{l} AssocT1mx\left(mo\right) \\ \sum_{ptype} T1tot\left(mo, ptype\right) \end{array} \right\} - AssocT1N\left(mo\right). \quad (A.20)$$

The numbers of inexperienced and experienced first-tour active-duty pilots in active-duty squadrons is then

$$ActT1N\left(mo\right) = \sum_{ptype} InexPlt\left(mo, ptype\right) - AssocT1N\left(mo\right). \quad (A.21)$$

$$ActT1X\left(mo\right) = \sum_{ptype} \left[T1tot\left(mo, ptype\right) - InexPlt\left(mo, ptype\right) \right] \\ - AssocT1X\left(mo\right). \quad (A.22)$$

Pilots in a Second or Subsequent Tour. The model needs pilots in a second or subsequent tour for reserve-component squadrons, active-duty squadrons, and alternative tours in the primary mission aircraft (mostly as FTU instructors). The number of pilots needed for reserve-component units is $AssocT2mx$ (*mo*), while the number needed for alternative tours is $AuthAlt2$ (*mo*).

The number of billets in active-duty squadrons that second-tour pilots are eligible to fill, which we denote by $AD2billets$ (*mo*), must satisfy the following conditions. First, there is a billet for any API-1 billets that first-tour pilots do not fill, plus the API-6 BMC billets. There may, of course, be more first-tour pilots than there are API-1 billets, but we do not allow the "extra" first-tour pilots to fill API-6 billets.

Second, the nominal, or "book," experience level must equal or exceed a user-specified target, which we denote by $NomExp$ (*mo*). This is equivalent to requiring enough experienced pilots to fill all the API-6 billets plus a specified fraction (the target experience level) of the authorized API-1 billets.[11] The book experience level is defined as

---

[11] The vast majority of second-tour pilots are experienced in the formal sense of having accumulated enough flying hours in the primary mission aircraft. If experienced pilots are not

$$BkExpLvl\left(mo\right) = \frac{ActT1X\left(mo\right) + AD2billets\left(mo\right) - BMC6\left(mo\right)}{CMRI\left(mo\right)}.$$

Then, the condition that book experience level equal or exceed a target is $BkExpLvl\left(mo\right) \geq NomExp\left(mo\right)$.

This expression imposes another condition on the required number of second- and subsequent-tour pilots, which, when combined with the first conditions, yields

$AD2billets\left(mo\right) =$

$$\text{MAX}\left\{\begin{array}{l} BMC6\left(mo\right) \\ BMC6\left(mo\right) + CMRI\left(mo\right) - \left[ActT1N\left(mo\right) + ActT1X\left(mo\right)\right] \\ BMC6\left(mo\right) + NomExp\left(mo\right) \times CMRI\left(mo\right) - ActT1X\left(mo\right) \end{array}\right\}. \quad \text{(A.23)}$$

We add the second-tour pilots in reserve-component squadrons and alternative tours to obtain the total billets available for second-tour pilots:

$$T2billets\left(mo\right) = AD2billets\left(mo\right) + AssocT2mx\left(mo\right) + AuthAlt2\left(mo\right). \quad \text{(A.24)}$$

The model can fill as many of these billets as it can from the pool of pilots awaiting a second operational tour. If all pilots in the pool have been assigned and there are still billets to fill, the model simply assumes enough pilots are available who have already had a second tour. Thus, the number of pilots moving from the pool into a second tour in month mo will be

$$Entry2\left(mo\right) = \text{MAX}\left\{0, \text{MIN}\left\{\begin{array}{l} T2billets - \sum_{tos=2}^{T2len} Tour2\left(mo, tos\right) \\ Pool\left(mo-1\right) + EntPool\left(mo\right) \end{array}\right\}\right\} \quad \text{(A.25)}$$

available, the Air Force Personnel Command may assign pilots that have not accumulated enough hours, expecting (hoping) that they can become experienced quickly. Often that is true, since these pilots generally will have flown a lot in other fighter aircraft.

This leaves the following number of second-tour billets to be filled by pilots in a third (or later) tour:

$$T3for2(mo) = MAX \begin{cases} 0 \\ T2billets(mo) - T2tot(mo) \end{cases}. \tag{A.26}$$

From Equation A.16 and the expression for $Temp2$ (mo, tos),

$$T2tot(mo) = Entry2(mo) + \sum_{tos=2}^{T2len} Tour2(mo, tos). $$

It may be necessary to overman active-duty squadrons—i.e., assign more pilots than are authorized—to achieve the desired book experience level. It is useful to define the following measure of the degree of overmanning of the active-duty squadrons. We calculate the total pilots assigned to operational squadrons as

$$\begin{aligned} ADassign(mo) &= ActT1N(mo) + ActT1X(mo) \\ &+ T2tot(mo) + T3for2(mo) - AssocT2mx(mo) \\ &- AuthAlt2(mo) + CMR6(mo). \end{aligned} \tag{A.27}$$

Then, the manning level for operational squadrons will be

$$ManLvl(mo) = \frac{ADassign(mo)}{\left[CMR1(mo) + CMR6(mo) + BMC6(mo)\right]}. \tag{A.28}$$

Cross-Sectional Constraints on Flying Hours

Only one loose end remains. To apply the equations given above, we need to know HCM (mo), the hours flown each month by inexperienced pilots. But HCM (mo) depends, in part, on the manning of the operational squadrons, which is calculated with the above equations. This section describes how the model calculates HCM (mo) from the quantities given earlier (including the manning of the operational squadrons) and how the model iterates to converge on the correct value of *HCM* (*mo*).

The Need to Allocate Flying Hours

According to RAP guidance, it is desirable for inexperienced CMR pilots to fly at one rate, experienced CMR pilots to fly at a slightly lower rate, and BMC pilots (who we assume are all experienced) to fly at a third, still lower rate.

But most flying in fighter squadrons is done in formations of two or four aircraft, and not all pilots are qualified to fly in every position in a formation. A two-ship formation has one two-ship flight lead position and one wing position. A four-ship formation has one four-ship flight lead position, one two-ship flight lead position, and two wing positions. If the squadron has too many pilots qualified to fly only as wingmen, none will be able to fly as much as a pilot qualified as a two- or four-ship flight lead will.

Qualification Versus Experience

Historically, the relationship between experience status and pilot qualifications has been fairly stable. Pilots qualified to fly only in the wing position or as two-ship flight leads are almost all inexperienced. The great majority of four-ship flight leads are experienced. We have taken advantage of this relationship in our past work to allocate flying hours based on the fraction of pilots that are experienced.[12] This has been important because data on pilot qualification are available only from the squadron to which the pilot is assigned (the squadron leadership also determines qualifications). By contrast, a pilot's accumulated flying hours, and therefore his experience status, are available from the Air Force Personnel file.

Simulators may—or may not—change the relationship between experience and qualifications. There is a proposal to count simulated flying toward experience, allowing a pilot to be considered experienced

---

[12] See, for example, Taylor, Moore, and Roll, 2000; Taylor et al., 2002; and Bigelow et al., 2003.

after just 400 hours of flying in the actual aircraft instead of the current 500 hours. (It might require 100 simulator hours—or more, or less—to make up the difference. That question has not been decided.) But each squadron commander decides for himself when a pilot under his command is ready to upgrade to flight lead, and there is no guarantee that a squadron commander will think time spent in a simulator will accelerate a pilot's acquisition of flight leader skills.

Thus we added a "qualification" milestone to a pilot's first tour that acts exactly like the "experience" milestone, but this milestone can occur after a pilot has accumulated a different number of flying hours. Experience is still the milestone used to decide when a pilot has completed his first operational tour. The new milestone, qualification, is used to allocate flying hours:

$FH2Q$ (*mo, ptype*) = The number of flying hours a pilot of type *ptype* must accumulate during his first operational tour to be deemed qualified.

Earlier, we introduced the variables HCM (*mo*) for the number of hours each inexperienced pilot flies in each month. We now take this same variable to denote the number of hours each unqualified pilot flies in each month. We anticipate that the experience milestone will always precede the qualification milestone—i.e., $FH2E$ (*mo, ptype*) ≤ $FH2Q$ (*mo, ptype*)—so every unqualified pilot will be inexperienced, and HCM (*mo*) will still be the hours each inexperienced pilot flies.

In our model, only pilots in a first operational tour can be unqualified, and these will be the pilots who have not been on station long enough to have accumulated $FH2Q$ (*mo, ptype*) flying hours. Using the same approach as with experience in Equation A.3 and starting at any month *mo*, we count back, accumulating the hours unqualified pilots flew in previous months until we reach a month in which the accumulated number of hours finally equals or exceeds $FH2Q$ (*mo, ptype*). Pilots who started the tour before then must be qualified. Pilots who started after then must be unqualified. Pilots who started at that time may or may not be qualified. In this case, the weighting is defined as follows:

If $\sum\limits_{j=1}^{tos} HCM\left(mo - j + 1\right) < FH2Q\left(mo, ptype\right)$, then

$$QWt\left(mo, ptype, tos\right) = 1.$$

If $\sum\limits_{j=1}^{tos-1} HCM\left(mo - j + 1\right) < FH2Q\left(mo, ptype\right)$ and

if $\sum\limits_{j=1}^{tos} HCM\left(mo - j + 1\right) \geq FH2Q\left(mo, ptype\right)$, then

$$QWt\left(mo, ptype, tos\right) = \frac{FH2Q\left(mo, ptype\right) - \sum\limits_{j=1}^{tos-1} HCM\left(mo - j + 1\right)}{HCM\left(mo - tos + 1\right)}.$$

If $\sum\limits_{j=1}^{tos-1} HCM\left(mo - j + 1\right) \geq FH2Q\left(mo, ptype\right)$, then

$$QWt\left(mo, ptype, tos\right) = 0.$$

$$(A.29)$$

Then, the number of unqualified pilots of type *ptype* is

$$UnqualPlt\left(mo, ptype\right) = \sum\limits_{tos=1}^{tosmax} QWt\left(mo, ptype, tos\right)$$

$$\times Tour1\left(mo, ptype, tos\right).$$

$$(A.30)$$

The next few quantities are calculated and displayed as part of the output used for analysis. They are not used in subsequent calculations. The time to qualification, if pilots become qualified, is

$$TTQ\left(mo, ptype\right) = \sum\limits_{tos=1}^{tosmax} QWt\left(mo, ptype, tos\right).$$

$$(A.31)$$

Pilots can finish a first tour without becoming qualified only if QWt (mo, ptype, tosmax) = 1. The number of such pilots is

If $QWt\left(mo,\ ptype, tosmax\right) < 1$, then

$$NeverQual\left(mo + 1,\ ptype\right) = 0.$$

If $QWt\left(mo,\ ptype, tosmax\right) = 1$, then

$$NeverQual\left(mo + 1,\ ptype\right) = Tour1\left(mo,\ ptype, tosmax\right).$$

(A.32)

The number of pilots that become qualified in month mo is

$$Qual\left(mo,\ ptype\right) = UnqualPlt\left(mo - 1,\ ptype\right)$$
$$- UnqualPlt(mo,\ ptype) + Entry1(mo,\ ptype)$$
$$- NeverQual\left(mo,\ ptype\right).$$

(A.33)

We assumed that experienced and inexperienced first-tour pilots were in the same proportions in reserve-component squadrons as in active-duty squadrons. We make an analogous assumption about qualified versus unqualified first-tour pilots. Thus, if $AssocT1U$ (mo) and $AssocT1Q$ (mo) are the numbers of inexperienced and experienced first-tour pilots in reserve-component squadrons,

$$AssocT1U\left(mo\right) = \frac{\sum\limits_{ptype} UnqualPlt\left(mo,\ ptype\right)}{\sum\limits_{ptype} T1tot\left(mo,\ ptype\right)}$$
$$\times \mathrm{MIN}\left\{\begin{array}{l} AssocT1mx\left(mo\right) \\ \sum\limits_{ptype} T1tot\left(mo,\ ptype\right) \end{array}\right\}$$

(A.34)

and

$$AssocT1Q\left(mo\right) = MIN \left\{ \begin{array}{l} AssocT1mx\left(mo\right) \\ \sum\limits_{ptype} T1tot\left(mo, ptype\right) \end{array} \right\}$$

$$- AssocT1U\left(mo\right). \tag{A.35}$$

The number of unqualified and qualified first-tour pilots in active-duty squadrons is

$$ActT1U\left(mo\right) = \sum\limits_{ptype} UnqualPlt\left(mo, ptype\right) - AssocT1U\left(mo\right) \tag{A.36}$$

and

$$ActT1Q\left(mo\right) = \sum\limits_{ptype} \left[T1tot\left(mo, ptype\right) - UnqualPlt\left(mo, ptype\right) \right]$$

$$- AssocT1Q\left(mo\right). \tag{A.37}$$

Allocation of Flying Hours in Active-Duty Squadrons

Hours flown in active-duty squadrons must be allocated among unqualified CMR pilots, qualified CMR pilots, and BMC pilots (whom we assume are all qualified). Equation A.36 provides the number of unqualified first-tour CMR pilots in active-duty squadrons. We must apportion the remaining pilots in second or third tours, $AD2or3tot$ (mo), between CMR and BMC.

We calculate the minimum number of second- or subsequent-tour pilots that must be CMR to be

$$CMRmin\left(mo\right) = CMR6\left(mo\right)$$

$$+ MAX\left\{ 0, CMR1\left(mo\right) - ActT1U\left(mo\right) - ActT1Q\left(mo\right) \right\}.$$

The minimum number of BMC pilots is $BMC6$ (mo). The excess of second- and subsequent-tour pilots over these minima is

$$ADexcess(mo) = ADassign(mo) - CMRmin(mo) - BMC6(mo)$$

The model assigns a fixed fraction *PlusBMCfrac* of the excess to fly at the BMC rate and the complementary fraction 1 − *PlusBMCfrac* to fly at the qualified CMR rate. Thus, the active-duty squadron billets that are available to second-tour pilots are apportioned between CMR and BMC flying rates as follows:

$$AD2CMRI(mo) = CMRmin(mo) - CMR6(mo)$$
$$+ (1 - PlusBMCfrac) \times ADexcess(mo) \tag{A.38}$$

and

$$AD2BMC6(mo) = BMC6(mo) + PlusBMCfrac \times ADexcess(mo). \tag{A.39}$$

Hours that Unqualified Pilots in Active-Duty Squadrons Fly

We estimate the total hours that aircraft in active-duty squadrons fly from the number of aircraft, their utilization rates (sorties flown per aircraft per month), and the length of each sortie. Let

$UTE(mo, AD)$ = Sorties per active-duty aircraft in month mo

ASD = Average sortie duration in hours.

Then, the hours flown by all pilots in active-duty squadrons is

$$FHtot(mo, AD) = UTE(mo, AD) \times ASD \times \sum_{sq \in AD} PAA(mo, sq). \tag{A.40}$$

Our method for allocating these hours to the three categories of pilots is rather unintuitive, but the results it yields have passed the scrutiny of highly experienced fighter pilots.

In the model, the user specifies the nominal number of sorties per month flown by BMC pilots, denoted by *SCMperBMC*. Almost always, the squadron will generate enough sorties to keep this number lower than the average number of sorties per active-duty pilot; if not, BMC flying is reduced to the average. Thus, the hours BMC pilots fly are given by

$$FHbmc\left(mo, AD\right) = AD2BMC6\left(mo\right) \times MIN \left\{ \begin{bmatrix} ASD \times SCMperBMC \\ \dfrac{FHtot\left(mo, AD\right)}{ADassign\left(mo\right)} \end{bmatrix} \right\}.$$

(A.41)

The model uses two rules to allocate flying hours to unqualified pilots in active-duty squadrons. First, neither BMC pilots nor unqualified pilots may fly at a higher rate than qualified CMR pilots. Second, flying by unqualified pilots is limited by the requirement for a pilot qualified as flight lead to supervise most of their sorties.[13] We will address the second rule first.

Define the actual qualification level as the ratio of qualified API-1 pilots to total API-1 pilots:

$$Quallvl\left(mo\right) = \frac{ActT1Q\left(mo\right) + AD2CMR1\left(mo\right)}{ActT1U\left(mo\right) + ActT1Q\left(mo\right) + AD2CMR1\left(mo\right)}.$$

(A.42)

Then, unqualified pilots may not fly more than the following fraction of total flying hours:

$$Ifrac\left(mo\right) = MIN \left\{ \begin{array}{l} Mxfrac \\ \begin{bmatrix} Wgfrac1 \times \left(\dfrac{Lev2 - Quallvl\left(mo\right)}{Lev2 - Lev1} \right) \\ + Wgfrac2 \times \left(\dfrac{Quallvl\left(mo\right) - Lev1}{Lev2 - Lev1} \right) \end{bmatrix} \end{array} \right\}.$$

(A.43)

---

[13] The Air Force measures a fighter pilot's experience in flying hours, but expresses his training requirements in sorties. In the model we can use the terms almost interchangeably, because flying hours equal sorties multiplied by a constant (ASD).

This is simply a linear interpolation between a fraction *Wgfrac1* at *Quallvl* = *Lev1* and a fraction *Wgfrac2* at *Explvl* = *Lev2*, truncated at a maximum fraction of *Mxfrac*. For this fraction, we have been using 0.377 of the total sorties and flying hours when the experience level is 60 percent, increasing that to 0.46 when the experience level drops to 40 percent. We never allow inexperienced pilots in active-duty squadrons to fly more than a fraction 0.46 of the total sorties and hours flown in active-duty squadrons.

The model assumes BMC pilots will fly a user-specified fraction of these hours, denoted by *OHfrac*, as wingmen. These hours are taken from the portion *Ifrac* (*mo*) of flying hours that might otherwise be available for inexperienced pilots. We thus define one limit on the total hours inexperienced pilots can fly to be

$$
\begin{aligned}
FHunqual1\left(mo\right) = {} & Ifrac\left(mo\right) \times FHtot\left(mo, AD\right) \\
& - OHfrac \times FHbmc\left(mo, AD\right).
\end{aligned}
$$

(A.44)

Another rule for determining how many sorties and hours an unqualified pilot may fly starts with calculating total hours all CMR pilots fly, whether qualified or unqualified. Unqualified pilots may not fly more than a pro rata share of the hours allocated to all pilots flying at the CMR rate:

$$
FHunqual2\left(mo\right) = \frac{\left[FHtot\left(mo, AD\right) - FHbmc\left(mo, AD\right)\right] \times ActT1U\left(mo\right)}{ActT1U\left(mo\right) + ActT1Q\left(mo\right) + AD2CMR1\left(mo\right) + CMR6\left(mo\right)}.
$$

(A.45)

We then use the supply of flying hours for inexperienced pilots as the minimum of the two quantities calculated in Equations A.44 and A.45:

$$
FHunqual\left(mo, AD\right) = MIN\left\{FHunqual1\left(mo\right), FHunqual2\left(mo\right)\right\}. \quad \text{(A.46)}
$$

Hours that Unqualified Pilots in Reserve-Component Squadrons Fly

The calculation of flying hours is simpler for reserve-component squadrons than for active-duty squadrons. Let

UTE (*mo*, *RC*) = Sorties per reserve-component aircraft in month *mo*.

As described earlier, we fence off some of the aircraft in each reserve-component squadron for the use of reserve and guard pilots, leaving the remainder for active pilots. This is, of course, equivalent to allocating a fixed percentage of a squadron's flying hours to active pilots. Current policy does not provide for any training of active pilots in reserve-component squadrons, which corresponds to fencing 100 percent of reserve-component aircraft. We investigated an alternative of fencing 15 aircraft in each reserve-component squadron, which would leave three aircraft in an 18-PAA squadron or nine in a 24-PAA squadron to provide training for active pilots. Let

FENCE (*sq*) = Aircraft in squadron *sq* withheld from use by active-duty pilots (fenced); applies only to squadrons *sq* ∈ *RC*.

Given these quantities, we calculate hours active pilots fly in active and reserve squadrons as:

$$FHtot\left(mo, RC\right) = UTE\left(mo, RC\right) \times ASD$$
$$\times \sum_{sq \in RC} MAX \left\{ \begin{array}{l} 0 \\ PAA\left(mo, sq\right) - FENCE\left(sq\right) \end{array} \right\}. \qquad (A.47)$$

We assume that every active-duty pilot in a reserve-component squadron flies the same number of hours. Thus,

$$FHunqual\left(mo, RC\right) = \frac{FHtot\left(mo, RC\right) \times AssocT1U\left(mo\right)}{AssocT1U\left(mo\right) + AssocT1Q\left(mo\right) + AssocT2mx\left(mo\right)}.$$
$$(A.48)$$

Hours per Month Flown by Each Unqualified Pilot

Recall that these calculations require a value for HCM (mo).[14] We are now in a position to recalculate the hours each unqualified pilot flies:

$$HCMrev\left(mo\right) = \frac{FHunqual\left(mo, AD\right) + FHunqual\left(mo, RC\right)}{\displaystyle\sum_{ptype} UnqualPlt\left(mo, ptype\right)}. \qquad (A.49)$$

$HCMrev$ (mo) will equal HCM (mo) only if we began with the correct value. This provides a way to adjust HCM (mo) iteratively to converge on the correct value.

It will be helpful at this point to introduce an index for counting these iterations. Let HCM_k (mo) and $HCMrev_k$ (mo) be the values of HCM (mo) and $HCMrev$ (mo) at the start and end of iteration k, respectively (at the start of iteration 1, we arbitrarily set HCM_1 (mo) = 15).

Always making the simplest possible guess for HCM_{k+1} (mo)—that HCM_{k+1} (mo) = $HCMrev_k$ (mo)—nearly always results in converging iterations. We have found, however, that successive values of HCM_k (mo) chosen in this way sometimes oscillate. If

$$\Delta_k\left(mo\right) = HCMrev_k\left(mo\right) - HCM_k\left(mo\right), \qquad (A.50)$$

HCM_k (mo) is oscillating if Δ_k (mo) × Δ_{k-1} (mo) < 0 (that is, changing sign from one iteration to the next).

If $HCMrev$ is treated as a function of HCM, say $HCMrev$ = $f(HCM)$, the objective is to find a fixed point of this function, HCM = f (HCM). Plotting successive pairs $(HCMrev_k, HCM_k)$, as shown in Figure A.1, leads to a pair that lies on the diagonal. Two successive iterations in which Δ_k changes signs yield two points on this function that lie on opposite sides of the diagonal, and a reasonable guess for HCM_{k+1} (the next trial value of HCM) is the value of HCM at which the line between the successive plotted values intersects the diagonal.

Thus, at iteration $k+1$, we select HCM_{k+1} (mo) as follows:

---

[14] At that point we called it hours per *inexperienced* pilot per month. Since we are assuming every inexperienced pilot is also unqualified, the two values will be the same.

Figure A.1
Selection of Next Trial Value When HCM Is Oscillating

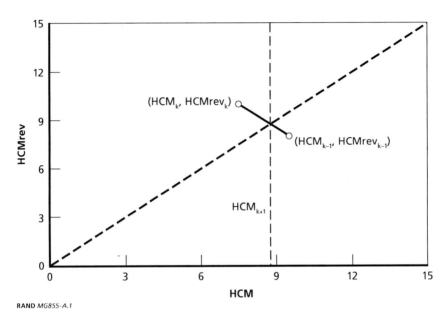

RAND MG855-A.1

If $\Delta_{k-1}(mo) \times \Delta_k(mo) > 0$, then

$$HCM_{k+1}(mo) = HCMrev_k(mo).$$

If $\Delta_{k-1}(mo) \times \Delta_k(mo) < 0$, then

$$HCM_{k+1}(mo) = \frac{\Delta_{k-1}(mo) \times HCM_k(mo) - \Delta_k(mo) \times HCM_{k-1}(mo)}{\Delta_{k-1}(mo) - \Delta_k(mo)}.$$

(A.51)

There is no need to consider $\Delta_k(mo) \times \Delta_{k-1}(mo) = 0$, for this case means that the process has converged.

We have not proved that this guess guarantees convergence, but as of this writing, it has yet to fail.

Trading Inputs for Outputs

For most uses, we provide the model with the inputs listed in Table A.1, and the model calculates, among other things, the outputs listed in Table A.2.

Sometimes, however, we want to establish a target value for one of the outputs and calculate the value of one of the inputs that is needed to achieve that target. We have constructed a variant of the model that does exactly this: We specify the desired manning level in each month—*ManLvl (mo)*, from Equation A.28—and the model finds the monthly entry rate of UPT graduates—*Entry1 (mo, UPT)*—that achieves that manning level.

This feat of trading inputs for outputs takes brute force. At each month, we pick a trial value for entry rate *Entry1 (mo, UPT)* and calculate all the outputs for that month, including *ManLvl (mo)*. The manning level cannot decrease as the entry rate increases. Therefore, we must reduce the entry rate if the manning level exceeds the target for that month or must increase the entry rate if the manning level falls short of the target.

We use the method of binary search, described earlier, to zero in on the correct entry rate. This scheme will work whenever there are

1. a single input quantity to vary in each month
2. a single target to achieve in each month
3. a way to tell unambiguously whether the input quantity is too high or too low.

See Chapter Three for some illustrative results using this feature.

Table A.1
Parameters Usually Provided as Model Inputs

Input Variable	Description
$FH2E\ (mo,\ UPT)$	Number of flying hours a UPT graduate must accumulate in his first operational tour to be deemed experienced
$FH2E\ (mo,\ FAIP)$	Number of flying hours a FAIP must accumulate in his first operational tour to be deemed experienced
$FH2Q\ (mo,\ UPT)$	Number of flying hours a UPT graduate must accumulate in his first operational tour to be deemed qualified
$FH2Q\ (mo,\ FAIP)$	Number of flying hours a FAIP must accumulate in his first operational tour to be deemed qualified
$Entry1\ (mo,\ UPT)$	Number of UPT graduates entering their first operational tour in month mo
$Entry1\ (mo,\ FAIP)$	Number of FAIPs entering their first operational tour in month mo
$tosmin$	Minimum length, in months, of a pilot's first operational tour
$tosmax$	Maximum length, in months, of a pilot's first operational tour
$T2len$	Length of a pilot's second tour
$DepEarly1\ (mo,\ ptype)$	Number of first-tour pilots to depart early
$TosEarly1\ (tos)$	Tos window for early departure of first-tour pilots
$DepEarly2\ (mo,\ ptype)$	Number of second-tour pilots to depart early
$TosEarly2\ (tos)$	Tos window for early departure of second-tour pilots
$PAA\ (mo,\ sq)$	Number of aircraft authorized for squadron, sq, in month, mo
$AD,\ RC$	Sets of active versus reserve squadrons
$FENCE\ (sq)$	Aircraft in squadron, sq, withheld from use by active-duty pilots (fenced); applies only to reserve-component squadrons
$UTE\ (mo,\ AD)$	Number of sorties flown per month (utilization rate) by aircraft in active-duty squadrons

Table A.1—Continued

Input Variable	Description
$UTE\ (mo,\ RC)$	Number of sorties flown per month (utilization rate) by aircraft in reserve-component squadrons
ASD	Average sortie duration in hours (assumed the same for active-duty and reserve-component squadrons)
$CR\ (mo,\ CMR)$	Crew ratios used for computing authorized CMR manning in active-duty squadrons
$CR\ (mo,\ BMC)$	Crew ratios used for computing authorized BMC manning in active-duty squadrons
$CMR6\ (mo)$	Authorized API-6 pilots who fly at CMR rates, not included in the authorizations calculated using crew ratios; applies only to active-duty squadrons
$AssocT1mx\ (mo)$	Active-duty pilots in a first operational tour authorized for reserve-component squadrons in month mo
$AssocT2mx\ (mo)$	Active-duty pilots in a second (or subsequent) operational tour authorized for reserve-component squadrons in month mo
$AuthAlt2\ (mo)$	Active-duty pilots authorized for an alternative second tour, usually as FTU instructor
$NomExp\ (mo)$	Smallest nominal or book experience level active-duty squadrons are permitted to have
$PlusMBCfrac$	Fraction of pilots in excess of authorized that are assigned to BMC billets
$SCMperBMC$	Nominal sorties per month flown by a BMC pilot
$Mxfrac$ $Wgfrac1$ $Wgfrac2$ $Lev1$ $Lev2$	Parameters used in calculating the maximum fraction of sorties and flying hours that inexperienced pilots in an active-duty squadron are eligible to fly
$OHfrac$	Fraction of BMC sorties and hours that encroach on inexperienced-eligible sorties and flying hours

Table A.2
Selected Parameters Usually Produced as Model Outputs

Output Variable	Description
$TTE\,(mo,\,UPT)$	Number of months since entry for pilots becoming experienced in month *mo*. It is a warning sign when this exceeds *tosmin*. But when it exceeds *tosmax*, pilots are exiting their first operational tour without becoming experienced. In practice, this happens only for UPT graduates.
$TTE\,(mo,\,FAIP)$	Number of months since entry for pilots becoming experienced in month *mo*. It is a warning sign when this exceeds *tosmin*. In practice, this exceeds *tosmax* only for UPT graduates.
$SCM\,(mo) = \dfrac{HCM\,(mo)}{ASD}$	Sorties flown by each inexperienced pilot in month *mo*
$BkExpLvl\,(mo)$	Book experience level actually achieved in active-duty squadrons
$QualLvl\,(mo)$	Actual qualification level of active-duty squadrons
$ManLvl\,(mo)$	Manning level of active-duty squadrons
$\dfrac{ActT1N\,(mo)}{CMR1\,(mo)}$	Fraction of API-1 billets in active-duty squadrons occupied by inexperienced pilots. If this ratio becomes too large, the active-duty squadrons find it very difficult to perform their combat missions.
$\dfrac{ActT1Q\,(mo)}{CMR1\,(mo)}$	Fraction of API-1 billets in active-duty squadrons occupied by unqualified pilots. If this ratio becomes too large, the active-duty squadrons find it very difficult to perform their combat missions.
$Pool\,(mo)$	Number of pilots awaiting a second tour
$WillWait\,(mo)$	Months a pilot just entering the pool in month *mo* will wait for a second-tour assignment
$DidWait\,(mo)$	Months a pilot just leaving the pool in month *mo* did wait for a second-tour assignment

The 2005 Aircrew Review

The 2005 Aircrew Review was held at Bolling AFB, Maryland, in December 2005. The review was sponsored by Lt Gen Carrol H. "Howie" Chandler, Deputy Chief of Staff for Air, Space, and Information Operations, Plans and Requirements (HQ USAF/A3/A5), and was attended by key leaders of various Air Force organizations.[1] General Chandler presented a slide entitled, "Why an Aircrew Review?" and provided the following answers to that question:

- It was time to share the CSAF's vision.
- The last review had been over five years before.
- Significant challenges were emerging.
- Training systems must be prepared to meet tomorrow's needs.
- The rated force must be prepared for the future.

Three key issues were to be discussed at the meeting: aircrew production, aircrew training systems, and emerging nonflying rated requirements. These were to be addressed in light of the facts that

1. A "total force" aircrew is needed that is capable and viable.
2. Active-duty and air reserve components are *interdependent*.
3. The ability to operate in a joint environment is important.

---

[1] Including the vice commanders of ACC, AMC, USAFE, AFSPC, AETC; the commander of AFRC, and the Deputy Chief of Staff for Installations and Logistics—all three-star generals.

4. Decisions must be affordable—the Air Force must get the most "bang for the buck."

We presented an analysis to the group that recounted some of the history discussed in Chapter Two of this document, then made the argument as follows.

Programmed Pilot Production Exceeds the Active Force Absorption Capacity

We described the aircrew management challenge in the following way: Air Force leadership wanted to maintain pilot production at current levels—1,100 SUPT graduates per year with about 330 of them going to fighter aircraft—because the Air Force needs rated officers in a variety of flying *and* nonflying billets, some of which are now vacant. At the same time, the fighter aircraft force structure is decreasing and transitioning (to the F-22, for example), which affects how many new pilots can be absorbed and the opportunities for second tours. The Air Force needs to conduct operational training and to fill other rated positions in an environment of constrained resources, so the question is whether or not this is possible.

We first defined the three unit states described in Chapter Four. To recap,

- A *healthy unit* has 100 percent manning—it has the number of pilots it is authorized to have; no more and no less. At the same time, about 60 percent of the API-1 pilots in the unit meet the criterion for being deemed experienced. Finally, inexperienced pilots are able to fly the number of sorties per month required to maintain CMR.
- A *stressed unit* is overmanned (between 105 percent and 110 percent of the number of pilots it is authorized to have), but the overmanning is being monitored and controlled. About 45 percent of the API-1 pilots are experienced; the inexperienced pilots are able to maintain CMR flying rates but must struggle to do so.

- A *broken unit* is overmanned at 120 percent (and typically getting worse). Fewer than 40 percent of the API-1 pilots are experienced, and new pilots cannot become experienced in a 36-month tour.

Figure B.1 displays modeling results (as of December 2005) related to the health of fighter units that were shown at the review. Excursions with the dynamic model showed that maintaining healthy fighter units requires reducing production to the levels represented by the lowest line—that is, slightly above 250 per year until 2005, but decreasing gradually to 200 thereafter. Slightly higher rates could be maintained if stressed units were acceptable, but the production levels programmed at the time (hovering around 300 per year) would likely break the units. Thus, the Air Force's production of new fighter pilots to fill rated and nonrated positions is constrained by the capacity of the training system to absorb new pilots.

Figure B.1
Effects of Different Levels of Pilot Production

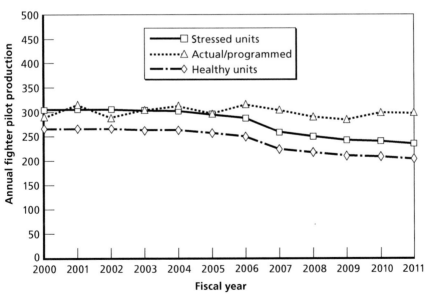

Training Consequences

The RAND presentation to the aircrew review included a series of slides similar to those in Chapter Three showing the effect of then-current policies on overmanning, TTE, and SCM and concluded that the following were the near-term consequences of current policies:

- The F-15C system will break (and remain broken indefinitely) with the programmed flow of new pilots.
- An F-16 system break can be prevented by flow reductions initiated in FY 2008, but the system remains under heavy stress throughout the FYDP.
- A-10 and F-15E systems will experience increasing stress during the FYDP—especially if CAF UTEs are not met or new pilots are diverted from other MDSs.

The RAND briefing offered several potential solutions to these problems. First, stated required numbers of rated personnel could be decreased, but this is unlikely, considering proposed increases in staffing levels for AOCs. This would also not solve the immediate problem of stressed units. Second, UTE rates for legacy aircraft could be increased. However, modeling showed that the UTE rates necessary to allow additional sorties to train more pilots would be infeasible. Third, the criteria for experience and other qualifications could be changed.[2] The final potential solution was to decrease the flow of new pilots to active units.

The results of model runs with reduced fighter pilot entries into active-duty units were presented to show that the reductions could prevent overmanning, increase SCM to acceptable levels, and ensure that new pilots could become experienced in a first tour.

---

[2] As described in Chapter Two, this was actually done for F-15Cs in 2006. While modeling showed that this improved the training environment slightly, it was not enough to prevent problems in the future.

Recommendations

The RAND briefing made four recommendations for immediate action:

- Reduce the flows of new pilots entering active F-15 and F-16 units quickly and in significant numbers by diverting some pilots to ANG/AFRC units or to other MDSs. Manage B-course output so that new-pilot flow *depends on* absorption capacity.
- Accelerate and expand active-associate initiatives to make better use of the available PMAI.
- Make the F/A-22 fully absorbable as quickly as feasible.
- Address the *full scope* of the fighter dilemma, including the inventory shortfall, operational unit health, and officer development.

In addition to these immediate recommendations, RAND recommended reexamination of nonflying rated position requirements. Analysis of rated positions that are currently unfilled might show that, for some, a pilot background is not actually required, other skills and experience from nonflying positions are adequate, or that non–active-duty personnel could fill a given position. It is also possible that closer scrutiny could show that some unfilled positions are no longer necessary, which would free rated personnel for flying positions.

RAND also recommended examining the potential use of total force assets from the ANG and ARC more closely. Although the three total force components operate as separate entities, the interdependence among them is long standing. This is illustrated by the fact that some 98 percent of the experienced guard and reserve fighter pilots were originally absorbed in active units. This means that the active absorption problem is definitely also a total force problem because the reserve components rely almost totally on highly experienced active pilots, separating after completing initial active-duty service commitments, to affiliate with guard or reserve units. If the active operational training systems break down to the point that this continuing flow of qualified pilots ceases, major changes will be required in the reserve components. It is in everyone's best interests to ensure that the active associate units become effective and that new units are developed as quickly as possible.

Indeed, beyond the FYDP, there will be even more-compelling reasons for the components to cooperate. As additional PMAI reductions occur, one factor certainly stands out: If the total force were to continue to rely almost solely on active units to absorb new fighter pilots, none of the components could be sustained in the future using current paradigms. It is also clear that it would be extremely difficult for the components to sustain themselves as independent entities in the future because that would essentially triple the current pipeline production and active absorption problems. The future total force fighter PMAI is quite unlikely to be large enough for the components to remain independent, operating as vertical stovepipes. The future would seem to argue for increasing consolidation of the components to ensure their survival. A fundamental objective of this consolidation would be to ensure that it will enable all three components to improve their circumstances.

Our presentation concluded with the following key components of a long-term solution to the existing (and emerging) aircrew management problems:

- Total force CAF aircrew requirements must be compatible (in terms of force structure and training resources) with a sustainable inventory.
- Adequate buffers and flexibility are needed in the aircrew flow system to accommodate surges, surprises, and program changes as they occur (approximately 85 percent of steady-state capacity).[3]
- Consolidation and increased efficiencies are necessary throughout the system.
- Compromises are essential, but consequences must be fully understood.

---

[3]   As in Taylor et al., 2002.

Working Group on Transformational Aircrew Management Initiatives for the 21st Century

Working Group Establishment

Lt Gen Carrol H. "Howie" Chandler, Deputy Chief of Staff for Air, Space, and Information Operations, Plans and Requirements (HQ USAF/A3/A5) established the TAMI 21 working group in August 2006. The focus areas for the working group were to be Air Force operators, the definition of experience, pilot production and absorption, and training. General Chandler's assessment of aircrew management goals included the following:

- revalidating all aircrew requirements: fill operator billets with operators, fill aircrew billets with aircrews
- expanding the use of the total force in all aspects of training, operations, and staff
- sizing the aircrew force to ensure that its members get the proper mix of flying, nonflying, and professional development during their careers
- ensuring a reliable flow of combat-ready aircrews to support a protracted long war, as well as other combat and support operations
- creating an efficient aircrew training pipeline that exploits the latest technologies and effective oversight mechanisms.

Maj Gen Dave Clary (Director of Air Force Operations under Lt Gen Chandler) was chairman of the working group and Lt Col Frank van Horn (Aircrew Management branch chief) was the action officer who managed the process. Members of the group included personnel from AFPC; AFSOC; A1PP; ANG/A1F, AFRC/A3T, AETC/A3R; ACC Air and Space Operations, Flight Management Branch; AMC/A3TF; and the RAND Corporation. Formal working group meetings were held 10–11 October 2006 (in Washington, D.C.), 14–16 November 2006 (at Randolph AFB, Texas), 28–30 November 2006 (Washington, D.C.), and 9–11 January 2007 (Washington, D.C.).

Working Group Charter

At the first TAMI 21 meeting, General Clary stated that the group's charter was to identify major aircrew management issues and develop a "10 Year Plan" to ensure a reliable flow of combat ready aircrews to the warfighters. General Chandler empowered the group to develop solution sets for major aircrew management issues the Air Force was then confronting to collect and assess background information, as required; and to present recommendations to senior leadership for action. The group framed the major issues as

- Requirements: How does an operator become qualified or credentialed to achieve "desired effects" for the warfighter?
- Experience: What does it mean? How do you get it?
- Sustainment and Absorption: Why do we produce aviators? Can we absorb more effectively? What is the effect of retention?
- Training: Are we training the right things the right way?

Working Group Recommendations

While TAMI 21 was chartered to look at aircrew management issues for all aircraft and rated personnel, work focused quickly on the fighter pilot community because of the problems raised in the 2005 Aircrew Review. By December 2006, the working group had developed a five-part approach to balancing the fighter-pilot production system by FY 2016—that is, ensuring that fighter units were not overmanned, that inexperienced pilots flew enough sorties per month, that experience levels were at or above 50 percent, and that first-tour pilots did not fill too large a proportion of API-1 slots. The components of the approach were

- Limit fighter pilot absorption in active-duty units to approximately 200 pilots per year.
- Sustain inventory requirements using alternative absorption options.
- Make full use of alternative manning options to reduce demand for active aircrews (AFRC/ANG, military-to-civilian, CEAs, other weapon systems, etc.).
- Constrain aircrew requirements (and their growth) to sustainable levels.
- Secure authority and provide necessary information for the director of Air Force Operations to manage aircrew production to meet inventory needs and unit health constraints.

Limiting fighter pilot absorption in active units to about 200 pilots per year is important because of the RAND modeling described in the main body of this monograph that shows the many problems that result if the Air Force attempts to absorb more. Since the Air Force anticipates the need for a larger inventory of personnel with fighter-type skills than can be maintained by absorbing only 200 fighter pilots per year, alternative sources must be considered.

One alternative absorption option is to create a new UAS career field and send 50 to 100 SUPT graduates per year to fly unmanned aircraft. UAS requirements are increasing, so these pilots would fill an emerging need. As UAS pilots, they would also develop the fighterlike skills (similar to those of the A-10) that the Air Force considers necessary to fill certain staff positions (such as rated positions at AOCs. Because of decreasing fighter aircraft infrastructure, it would be very unlikely that pilots who take the UAS route would be able to fly another fighter aircraft later in their careers.

Another alternative absorption option makes use of ANG and AFRC aircraft. By 2016, reserve units will own 36 percent of the total force PMAI—361 fighter aircraft compared to 632 in the active inventory. After completing FTU in their fighters, some new pilots could go to guard or reserve units for a first tour to become experienced pilots. After that tour, they would return to an active-duty unit. The working group concluded that the ARC could absorb between 50 and 75 pilots per year.

The third and fourth TAMI 21 recommendations address the need to clarify staff demands for active-duty fighter pilots. Further analysis could show that some of these positions could instead be filled by guard or reserve pilots, retired fighter pilots, or CEAs. Additionally, some of the emerging "requirements" for fighter pilots (in AOCs, for example) may not be necessary, and the growth of these requirements must be limited to sustainable levels.

The final TAMI 21 recommendation was to reaffirm the regulatory authority of the director of Air Force Operations to manage the production of rated officers to meet Air Force requirements and adjust production when necessary to ensure the health of a weapon system.

RAND Modeling

We used the dynamic model described in the body of this monograph to analyze how the many options the TAMI 21 working group considered would affect fighter pilots. The effect of implementing the working group's final recommendations was very positive for F-15Cs and F-16s: Manning levels remained below 105 percent, and SCM for inexperienced pilots remained between 8 and 10 from FY 2010 on. By implementing the recommendations, the Air Force could ensure the ability of fighter pilots to become experienced well before the end of a first operational tour and that the proportion of experienced pilots in units remained above 50 percent after 2010. Finally, the proportion of API-1 slots that first-tour pilots fill remains below 60 percent after FY 2010. In fact, the high experience levels shown by the model indicated the potential to increase absorption rates in the future.

TAMI 21 Implementation

The chairman of the TAMI 21 working group discussed the group's recommendations in a briefing with CSAF on December 18, 2006. The detailed recommendations were not approved, but continued discussions eventually led to the policy decisions described in Chapter Four.

References

Air Force Instruction, 11-103, *Aircraft Standard Utilization Rate Procedures*, November 15, 2004.

———, 11-412, *Aircrew Management*, April 25, 2005.

———, 11-2F-15, *Flying Operations,* Vol. I: *F-15 Aircrew Training*, January 18, 2007.

———, 65-503, *Financial Management: US Air Force Cost and Planning Factors*, February 4, 1994 (updated September 2008).

AirForce-magazine.com, "UAVs Are the Wave of the Future," in "Air Force Daily Report," February 21, 2008. As of April 9, 2009:
http://www.airforce-magazine.com/DRArchive/Pages/default.aspx

Anderegg, C. R., *Sierra Hotel: Flying Air Force Fighters in the Decade after Vietnam*, Washington, D.C.: U.S. Air Force History and Museums Program, 2001.

Ausink, John A., Richard S. Marken, Laura Miller, Thomas Manacapilli, William W. Taylor, and Michael R. Thirtle, *Assessing the Impact of Future Operations on Trainer Aircraft Requirements*, Santa Monica, Calif.: RAND Corporation, MG-348-AF, 2005. As of March 16, 2009:
http://www.rand.org/pubs/monographs/MG348/

Bigelow, James H., William W. Taylor, S. Craig Moore, and Brent Thomas, *Models of Operational Training in Fighter Squadrons*, Santa Monica, Calif.: RAND Corporation, MR-1701-AF, 2003. As of February 12, 2009:
http://www.rand.org/pubs/monograph_reports/MR1701/

Carney, Major, *Talking Paper on Rated Requirements vs. Inventory: 1 October 2007 Redline/Blueline*, talking paper, HQ USAF/A1PPR, copy provided February 25, 2008.

Chief, Rated Force Policy HQ USAF/A1PPR, email, October 5, 2007.

Clary, David (briefer), "Transformational Aircrew Management in the 21st Century," flight briefing to TAMI 21 Task Force members at Randolf Air Force Base, December 18, 2006.

Darnell, Lt Gen Daniel J., "Medium and High Altitude Unmanned Aircraft Assignments from SUPT and ENJJPT," unclassified message released from daadministrator@ptsc.pentagon.mil on September 17, 2008 (date time group: 171235Z SEP 08).

Drew, John G., Kristin F. Lynch, James M. Masters, Robert S. Tripp, and Charles Robert Roll, Jr., *Options for Meeting the Maintenance Demands of Active Associate Units*, Santa Monica, Calif.: RAND Corporation, MG-611-AF, 2008. As of February 12, 2009:
http://www.rand.org/pubs/monographs/MG611/

GAMS Development Corporation, General Algebraic Modeling System (GAMS) home page, Washington, D.C., undated. As of February 12, 2009:
http://www.gams.com/

Government Accountability Office, *DoD Aviator Positions: Training Requirements and Incentive Pay Could be Reduced*, GAO/NSIAD-97-60, February 1997.

————, *Military Personnel: DoD Needs to Address Long-Term Reserve Force Availability and Related Mobilization and Demobilization Issues*, Washington, D.C., GAO-04-1031, September 2004.

Headquarters AETC/FMATT, email, December 22, 2008.

Hoffman, Michael, "Gates: No More Cuts to Air Force Personnel," *Air Force Times*, June 11, 2008. As of February 12, 2009:
http://www.airforcetimes.com/news/2008/06/
airforce_drawdown_ends_060908w/

LaGrone, Sam, "Guard to Train RC-12 Pilots, Sensor Operators," *Air Force Times*, September 18, 2008. As of September 30, 2008:
http://www.airforcetimes.com/news/2008/09/airforce_guard_rc12_091708w/

Marken, Richard S., William W. Taylor, John A. Ausink, Lawrence M. Hanser, C. R. Anderegg, and Leslie Wickman, *Absorbing and Developing Qualified Fighter Pilots: The Role of the Advanced Simulator*, Santa Monica, Calif.: RAND Corporation, MG-597-AF, 2007. As of February 12, 2009:
http://www.rand.org/pubs/monographs/MG597/

Randolph, Monique, "Changes on the Horizon for Air Force Pilots," Air Force Link, May 15, 2007. As of March 16, 2008:
http://www.af.mil/news/story.asp?id=123054831

Taylor, William, "RAND White Paper No. 3: Improving Fighter Pilot Manning and Absorption," in Thie, et al., 2004, pp. 33–40.

Taylor, William W., S. Craig Moore, and C. Robert Roll, Jr., *The Air Force Pilot Shortage: A Crisis for Operational Units?* Santa Monica, Calif.: RAND Corporation, MR-1204-AF, 2000. As of February 12, 2009:
http://www.rand.org/pubs/monograph_reports/MR1204/

Taylor, William W., James H. Bigelow, S. Craig Moore, Leslie Wickman, Brent Thomas, and Richard S. Marken, *Absorbing Air Force Fighter Pilots: Parameters, Problems, and Policy Options*, Santa Monica, Calif.: RAND Corporation, MR-1550-AF, 2002. As of February 12, 2009:
http://www.rand.org/pubs/monograph_reports/MR1550/

Thie, Harry J., Raymond E. Conley, Henry A. Leonard, Megan Abbott, Eric V. Larson, K. Scott McMahon, Michael G. Shanley, Ronald E. Sortor, William W. Taylor, Stephen Dalzell, and Roland J. Yardley, *Past and Future: Insights for Reserve Component Use*, Santa Monica, Calif.: RAND Corporation, TR-140-OSD, 2004. As of February 12, 2009:
http://www.rand.org/pubs/technical_reports/TR140/

U.S. Air Force, A3O-AT, "Aircrew Management," PowerPoint briefing, July 9, 2008.

Wicke, Russell, *Rising Fuel Costs Tighten Air Force Belt*, Air Force Link: Official Web Site of the United States Air Force, September 8, 2006. As of November 2007:
http://www.af.mil/news/story.asp?id=123026679

Wynn, Michael W., Secretary of the Air Force, Testimony Before the House Armed Services Committee, October 24, 2007. As of August 1, 2008:
http://armedservices.house.gov/pdfs/FC102407/Wynne_Moseley_Testimony102407.pdf